Metallurgical Principles of Founding

Metallurgical Principles of Founding

V. Kondic, Ph.D.

Department of Industrial Metallurgy,
University of Birmingham

NEW YORK

AMERICAN ELSEVIER PUBLISHING COMPANY, INC.

© V. KONDIC 1968

First published 1968

American Elsevier Publishing Company, Inc.
52 Vanderbilt Avenue
New York, N.Y. 10017

Library of Congress Catalog Card Number: *68-24801*

PRINTED IN GREAT BRITAIN AT THE PITMAN PRESS, BATH

PREFACE

This book is intended for the student interested in founding for educational, vocational, professional or business reasons. With educational as well as pragmatic objectives in mind, an attempt has been made towards a single and unified rather than a general and comprehensive approach to founding. The approach selected is that of looking at founding in terms and ideas of applied science. This, rather than either the empirical or engineering approaches, was chosen because it was felt that the vast majority of students to whom the book is directed would have prior basic knowledge of physics and chemistry. (The level assumed in these subjects and in mathematics is that of the matriculation standard.) It follows immediately that the book deals with some but not all foundry activities. However, in order to make the book more readable, in many instances the engineering and empirical elements of founding have been added to the text. This is intended to give the book the necessary completeness as a single text, but for more adequate empirical or engineering treatment of founding the student should refer to separate reading.

Altogether, the scope and treatment of the subject matter of the book are aimed to provide an introductory metallurgical text for further studies of founding in breadth as well as in depth.

The book concerns the application of selected principles of metallurgical science to the processes of metal founding. The title therefore raises two questions: What is founding, and why is it necessary to isolate some metallurgical principles of founding into a separate textbook? In the introductory chapter of this book, the subject matter of founding is considered at greater length. For the benefit of the preface reader, however, a short answer could be given along the following lines. In the simplest terms, founding or casting of metals implies, firstly, the provision of a mould with a cavity which is a faithful reproduction of a part or the whole of the object to be made and, secondly, filling the mould cavity with the required metal. Such a process of making metal objects has played an important part, and in certain respects a key part, in the material existence of man from the beginnings of the metal age. Whilst in the early periods of history castings were used mainly for tools and ornaments, today castings enter into many phases of our industrial life, ranging from production machines and vehicles, on the one hand, to instruments and household goods on the other.

In writing the book, an attempt has been made to deal with an important educational problem of founding. The fact that such an educational problem exists at present is familiar to most teachers of foundry subjects, and could be stated in the following terms.

The treatment of founding in textbooks or in lectures at the majority of Technical Colleges or Universities today is similar to that of metallurgy treatment of about thirty years ago. Founding is lumped together as an omnibus subject: a mixture of science and technology, engineering and skilled practice, developments and production. Metallurgy today, however, has grown into two distinct disciplines—the science of metals based on physical and chemical sciences and metallurgical engineering, bringing in, in addition, the engineering sciences. Having been in contact with foundry education now for many years, the author has formed a firm opinion that treatment of founding as an educational discipline at the centres of higher education must develop along a similar line to that which has occurred with metallurgy. In other words it is necessary to consider applied science and engineering courses that enter into the field of founding as distinct and separate subjects but which together form a unified educational discipline—that of metal founding.

In line with such a development there is a need for separate texts dealing with various aspects of the different subjects that together make the discipline of metal founding. Among such subjects the two most important ones are certain branches of metallurgy and of engineering. This book on metallurgical principles of founding has therefore been written firstly to bring out those principles of the science of metallurgy that provide a foundation for studies of metal founding, and secondly to consider the application of such principles to current foundry practice. It is this combination of principles and their applications that distinguishes it from some of the available books in this field. In future, separate texts dealing with other aspects of metal founding will no doubt be necessary, but, for the present transitional period in foundry education, it is hoped that the present text may be of some help to students and teachers of founding.

This book, then, is an introductory text for students of metallurgy or engineering who may be taking founding as a subsidiary subject, or a first or second year text for students of founding. In the former case the book may be useful in pointing out some of the difficulties in applying pure science, while to the foundry technologist it may serve to emphasize the need for studies in pure science before dealing with their application. In either case the book will do more than justify its publication if it induces the student to develop an interest and to extend his knowledge of the fundamental aspects of metal founding. The idea of a casting process can be readily appreciated directly by visual observation, but the translation of an outline of an idea into a well-designed casting process and product requires an appreciation of a large number of concepts of fundamental and applied sciences. This book has been written as a guide book for a student who wishes to explore and understand this route.

In closing this preface I should like to express my deep gratitude to my numerous friends at the centres of education and in the foundry industry whose help with my own foundry education and with the publication of this book has been both generous and extensive. None of them, I feel sure, would object if I mention one in particular, Dr. W. T. Pell-Walpole, whose grasp of what is fundamental, and an ability to translate this into what is applied or practical, have been matched only by his extreme kindness and patience in teaching me the principles of metallurgy. My debt to my wife goes beyond what I could readily express in words, which are but a poor substitute for love.

V. K.

CONTENTS

PLATES

CHAPTER 1

THE FOUNDRY INDUSTRY

1.1 PAST AND PRESENT

Founding or metal casting is by definition any process of melting metals and pouring them into moulds in order to produce the required solid shapes. Such processes have been a human activity for well over 4,000 years. Metals and alloys (and more recently other materials) have been processed by different methods of founding and casting into a multitude of products. Sculpture, jewellery and tools are some of the typical products which were made as castings many centuries ago and in many instances are still produced in this manner. Machine, motor car, aircraft and rocket components are types of casting more representative of the present age.

In terms of the material and technological progress of man, founding can be taken as an index of the state of industrial development, or often as a barometer of the state of the economy, of a society. Metals are still the key materials of our industrial age, and founding is the initial and basic process of rendering metals, and frequently other materials, into useful objects. The usefulness of metals is based on some of their unique physical properties, mainly those of a combination of strength and ductility under the varied conditions of their use. The uniqueness of founding lies in the fact that in many cases it is the simplest and most economic and, in others, the only technically feasible method of obtaining a required solid shape. The relative technical simplicity with which complex shapes can be produced in most metals, and the economically competitive nature of the process when compared with alternative methods of manufacture, are the two most valuable characteristics of founding.

If we ignore artistic and social considerations, we can arrive at an assessment of the significance, merit or usefulness of a material product such as a casting from a purely industrial and technological point of view. A casting is normally used to meet some definite functional requirement in service, and its performance compared with that of possible alternative products can be assessed in terms of specific properties of the alloy in which the casting is made and the standards of such properties actually achieved in the casting. The degree of industrial usefulness can be obtained by evaluating the actual industrial performance of a product and relating this to the overall costs of its production. In some instances the properties required and/or the design and production economic factors are such that a clear cut decision can be made as to whether a casting is suitable, or more advantageous than the same solid shape produced by an alternative method of manufacture. In other cases a decision between several possible alternative processes for producing an article may require a searching and careful preliminary analysis. The metallurgical principles of assessment required for such evaluations of 'product quality index' are considered in this book, but more detailed production and engineering aspects are outside its scope, and are discussed in the books recommended for further reading.

1.2 STRUCTURE OF THE FOUNDRY INDUSTRY

The various components of the foundry industry of an industrialised country can be grouped in three distinct ways, on the basis of the composition of metals, the kinds of products made and the type of casting processes used.

1.2.1 Composition of alloys

Through the combination of several of their physical and chemical properties, certain metals and their alloys are more easily than others melted and converted into cast products of satisfactory quality. Cast or grey irons are an example of such alloys, where a combination of useful properties and cheapness of process and product explain the fact that the grey or cast iron foundry industry is, in terms of its economic significance, the leading branch of the foundry industry in most countries. Steel, malleable iron, copper and nickel, light alloys and zinc founding are other main branches of the foundry industry, and can be clearly distinguished in terms of the chemical composition of the castings produced. Such foundry divisions, in terms of metals and/or types of casting made in them, are of greater significance for the needs of trade and its organisation, research associations and individual industrial groupings than for the actual usage of the different casting processes or applications of the casting made. Least of all are they significant for the basic understanding of founding as a branch of applied science or engineering.

1.2.2 Products

The foundry industry can be divided into two large groups according to the type of product made, namely those producing ingots and those making shaped castings. Ingots, Fig. 1.1, are cast shapes, usually of simple geometry, which are mainly used for subsequent working to produce other shapes (sheet, strip, sections, bar, wire, etc.) by various processes of wrought metal manufacture (e.g., rolling, forging, extrusion and drawing). To a limited extent, ingots are used directly after machining (e.g., bearings), but more frequently simple shaped ingots are used as charge materials for melting in foundries. Ingot casting foundries are invariably attached to and part of either the metal fabrication or the metal extraction and refining industries.

Castings of an almost limitless variation of shape and design are produced in foundries which are usually described in terms of some of the characteristics of the process used—sand, gravity, precision and other types of foundry. The products of such foundries are referred to as shaped castings, implying

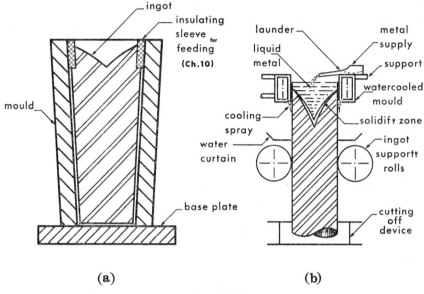

(a) **(b)**

FIGURE 1.1

(a) Batch ingot and its mould; one of many mould designs for casting slab steel ingots.

(b) Continuously cast ingot, mould and ancillary equipment; here again there are many machine and mould designs utilising different principles.

that the basic solid shape first produced is retained in the subsequent application of the casting. Minor changes in shape may be unavoidable (e.g., trimming, fettling and machining), and in some cases several castings can be welded together to produce the final required shape. The basic characteristic of shaped castings lies, however, in the fact that no process of plastic deformation or working is used to modify or alter their basic as-cast shape. Shaped castings can be equally satisfactorily defined by the characteristic that they have a certain specific as-cast structural character, which in turn controls many of the important properties of cast metals. The formation of the internal structure in castings, normally referred to as macro- or microstructure, its possible modification by controlling the solidification process or by subsequent heat treatment, and the general significance of the cast structure in relation to the properties of cast metals are among the major subjects discussed in the present book.

The essential problems in discussing shape, process and cast structure relationships in the processes of manufacture can be illustrated with reference to Fig. 1.2. The component typified by design (a) could be made as a casting, forging or pressing, but the cast structural character of the first would be generally distinct from that of the latter two products. Consequently some of the properties would be different, quite apart from other differences in economic factors of manufacture. With example of design in question, such simple issues as the actual weight of the component, or plasticity of the required alloy, may determine the nature of the most suitable manufacturing process. Very heavy objects must generally be cast and this is true of products

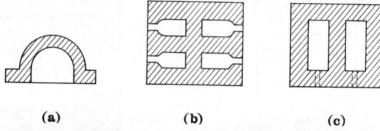

(a)   **(b)**   **(c)**

FIGURE 1.2 Cast shapes and the problem of formation of internal surfaces.

(a) An open shape suitable for making by a number of different methods.
(b) More enclosed internal surfaces restricting the number of shaping methods.
(c) Totally enclosed internal surfaces; shapes favourable for casting processes.

in alloys which are difficult either to machine or to deform plastically for shaping. With more complex shapes, including some enclosed surfaces such as example (b) in Fig. 1.2, the likelihood increases of the casting process being technically and economically the most appropriate choice, while the casting process is an obvious choice for design (c). In general, making moulds for a casting process is a technically simpler and more economic solution to producing complex shapes than the alternative methods of manufacture.

1.2.3   Processes

In the same way that some solid shapes can be produced by casting or by some alternative shaping method, many shapes can be produced by one of several casting processes. These processes can be grouped into two categories, depending basically either on the nature of the mould or on the manner of pouring liquid metal into and filling the moulds. From the industrial production point of view, these groupings imply the application of different machines for manipulating the moulds and pouring the metal. For the purpose of production it is generally easier to organise a foundry plant on a single process or group of similar types rather than one in which several distinct and different moulding and casting processes are combined. Hence the existence in the industry of sand (silica or other mineral) and metal mould foundries, where the nature of the moulds used forms the basis of classification, and of gravity, pressure and centrifugal foundries classified according to the method of pouring the molten metal into the moulds. Such clear distinctions do not always hold and in some foundries, particularly in sand foundries, use is made of certain features taken from different moulding processes. Such divisions are further discussed in §1.3.6.

Other terminology frequently encountered in connection with foundry processes includes the following. Light, medium and heavy foundries are those which produce shaped castings weighing up to 2 cwt, from 2 to 10 cwt and over 10 cwt respectively; jobbing foundries make a few off, of identical castings but in a relatively large variety of designs, while in repetition foundries castings of the same design are mass produced.

It is clear from this brief summary that the foundry industry is relatively complex in its structure: in the kinds of metals used, the types, quantity and size of castings made, and the variety of moulding and casting processes applied. Some of the more primitive methods of founding, using clay and stone moulds, still persist in certain parts of the world, in contrast to the existence of fully automated modern engineering casting plants in industrial countries. Foundry production units vary in size from those employing a few men to large ones employing several hundreds of men. Castings are now produced in well over 1,000 different casting alloys, and cast products are essential for running the transport, electrical and machine tool industries. Many castings are encountered in use in our daily domestic life, for example, taps, grates, and cooker parts. The economic significance of the foundry industry can be seen from the fact that about 25% of all the metal cast for the industrial needs in the world are in the form of shaped castings and most of the remainder is cast into ingots. Only a small fraction of the metals used are processed by powder, coating, or electroforming processes. In Great Britain, the annual output of shaped castings is at present about 5 million tons, and there are approximately a quarter of a million people employed in this branch of the industry.

1.3 FOUNDRY PROCESSES

1.3.1 Ingot casting

The general problem of ingot casting can be stated as follows: blocks of required size, geometry and properties have to be produced in various alloys so that they can subsequently be readily worked into various finished or semi-finished products of an appropriate quality and at an optimum production economy. An overall analysis of ingot casting, and similarly that of shaped castings and products of other processes of metal manufacture, involve considerations of metallurgical, engineering and production economic factors. The case of ingots used for subsequent shaping by various metal working processes will now be considered in more detail.

1.3.2 Ingots for subsequent working

Most of the industrially required products or semi-products made from ingots are of simple cross-sections—rounds, squares, tubes and other similar simple sections. These products are then further processed by forging, pressing and various joining and welding methods into a multitude of different engineering designs and constructions. Ingot casting and subsequent working and finishing processes are therefore components of an integrated system of production, and many of the problems arising at each stage of the process are closely inter-related to those of the preceding and subsequent stages.

For general industrial requirements, ingots can be produced by two distinct types of process, based on the application of batch and continuous casting principles respectively, Fig. 1.1. In either case the starting ingot size, shape and properties are aligned to the shaping processes that follow and to the working plants available.

In a batch process the mould, usually made in metal, is filled, the liquid metal is allowed to solidify, and after removing the ingot (or stripping the mould) the casting process is repeated. In a continuous process, the bottom of the mould is temporarily closed at the start of the casting process and then, as the ingot gradually solidifies, it is continuously withdrawn from the mould. After a given length of ingot is cast, the ingot can be removed and the process repeated (semi-continuous casting), or the pouring can be carried on without interruption and the required ingot lengths cut off below the mould (fully continuous process). Three distinct groups of problems—metallurgical, engineering and production engineering—can therefore be clearly distinguished.

Metallurgically, it is necessary to arrange and control the sequences of melting, pouring and solidification in such a way that an ingot is produced having structural and surface properties which are satisfactory or acceptable for the subsequent working processes, as well as for the final product requirements. At each stage of operation there are several process variables giving a wide range of alternative solutions for resolving all the different metallurgical problems which arise.

Engineering problems of ingot casting include such questions as the design and construction of the moulds and of the casting equipment and plant, together with those of plant maintenance and process controls. Production engineering problems deal with many and varied questions ranging from those of management and administration to those of work study and costing. The scope of problems and their solution encountered in engineering fields of ingot casting is as wide and as searching as that encountered in the metallurgical field of the process.

Many of the problems met in ingot casting cut across all three fields—metallurgy, engineering and production engineering. One such example is that of the selection of optimum ingot size and optimum properties in relation to the subsequent working processes and to the production economy. It is indeed the consideration of such problems which has led to the most striking developments in ingot casting processes in the industry. This has been particularly noticeable in the development of the continuous casting processes. It is clearly feasible to cast ingots very closely in size and geometry to that of the final products or semi-products. This in turn has considerable effect on the amount of subsequent working and therefore on the plant and economy of the working processes. Some examples of this category are: casting of hollow ingots for tube making, small ingot sections for strip rolling and double 'T' sections for rail rolling. The major metallurgical problems encountered in ingot casting are considered in Chapter 8, but the engineering and production engineering aspects are discussed in the recommendations for further reading.

A superficial consideration of ingot casting might lead one to suppose that such an apparently straightforward problem as that of casting metal blocks of simple geometry would by now have been fully mastered and understood in all its applied science and industrial implications. A close examination of the history of the processes used and of current knowledge in the field shows, however, that this is far from being so. The variety of alloys used, the wide range of possible casting methods to meet the demands of types and quality of products and the ever-changing economic problems of materials and of the productivity index, make the field of ingot casting a complex and challenging field of study. This is almost as true today as it was 100 years ago.

1.3.3 Other ingots

The problems arising in casting ingots used for machining or those used for remelting are similar in kind to those mentioned in the preceding paragraphs, the differences being mainly in the relative importance of various factors. For example, metallurgical problems are more dominant with ingots used for machining, whilst those of engineering are more relevant in casting of ingots for remelting.

1.3.4 Shaped castings

A familiar saying attributed to the skilled foundrymen is that 'whatever design an engineer can draw, a foundryman can make as a casting'. Whilst such a claim reflects something of the characteristic pride in the skills and art of founding, it is not a suitable statement on which to base a discussion of the knowledge of founding as a human activity. The present state of development of both the basic knowledge and the craft of founding is a result of interactions between practice and skill on the one hand and applications of engineering and metallurgical science on the other, as well as the consequence of sociological and economic forces in a changing industrialised society. In the subsequent paragraphs various currently used methods of moulding of shaped castings are summarised, with the object of presenting an overall and general picture of the present position. Various technical and economic factors which have contributed to present practice will first be reviewed and then some major characteristics of various processes considered. A more detailed discussion of the significance of various factors appears in later chapters of this book.

1.3.5 Technical and economic factors

Four major groups of factors are responsible for the way in which the use of different moulding and casting methods for making shaped castings has developed.

(*a*) *Alloy composition.* The melting temperature of an alloy may require specific types of mould material, while the design of the casting may require that the alloy should have good casting or founding properties. For example, alloys having high melting points cannot be readily cast into metal moulds to obtain shaped castings, and complexity of casting design imposes severe demands on the control of mould filling and of solidification, both of which are dependent on the alloy composition.

(*b*) *Casting weight and design.* Heavy castings (> 1 ton) can be made more readily in moulds made of compacted sand minerals than in metallic moulds. Also the technical difficulties of making shaped castings in metal moulds increase with increasing complexity of casting design.

(*c*) *Properties of the castings made.* Different types of mould impose their own characteristics on the various properties of the casting produced. A casting made in a sand mould will differ to a certain degree and in some specific properties from a casting of identical design made in a metal mould.

(*d*) *Economic factors.* The number of castings required, just a single one, or several thousands, may determine the economic choice of a casting process. Sand or mineral moulds are as a rule consumable, the mould being destroyed to remove the casting, whilst a metal mould can be used, to a degree, repeatedly. Similarly, the cost of various raw materials required for mould making, the speed of making consumable moulds in various moulding mixtures, the cost of mould making equipment and other production factors, all have to be considered in assessing the economic characteristics of different moulding methods or types of mould.

1.3.6 Moulding and casting methods

The basic requirements of a casting method for making shaped castings can be deduced by reference to Fig. 1.2. If one of the designs shown is to be made as a casting, then it is necessary to construct a mould in which a cavity is provided which corresponds to the design of the casting required. Some of the main methods which can be used for the making or construction of moulds will be briefly summarised.

(a) *Sand moulds.* In this method, a moulding mixture based on silica sand or some other mineral and moist clay is rammed on to a pre-made pattern (or part of a pattern) which is identical in basic shape with the casting required. The mould is usually made in parts, so that the pattern can be withdrawn and the parts of the mould assembled together to give the desired mould cavity as well as the necessary channels for filling it to obtain a satisfactory casting. A wheel type casting moulded and poured in this way is illustrated in Fig. 1.3. The variety of moulding mixtures which can be used, the variety of designs of patterns and of methods of construction of external and internal parts of moulds (cores), the range of equipment which can be used for mould making and the considerations involved in arranging such equipment into a production flow, all together make sand founding the most versatile, basic, representative and thought stimulating of all the moulding processes. The relative freedom in choice of alloys and of casting design is its main advantage, and difficulties in meeting the more exacting metallurgical and engineering specifications for some castings are the main limitations. Sand founding is a general term and covers a large number of process variations where either the nature of the moulding mixture, or specific conditions of the mould prior to pouring, or method of ramming the mould differs from that used in the conventional process (dry sand, CO_2 sand, loam, cement and compo sands, oil-core sand, 'fluid' sand, hand or machine moulding and so on).

(b) *Variants of sand moulds.* In some variants of sand moulding, whilst the basic principle of moulding is retained, a major departure in the technique is significant. For example, in shell moulding, Fig. 1.3b, the sand and a specific binder mixture (Chapter 8) are allowed to harden on a hot metal pattern, giving a shell which fastened to or glued together with other shell parts provides the required mould cavity. In another method, precision or lost wax moulding, Fig. 1.3d, a wax pattern is encrusted in a special mineral-binder mixture, which hardens sufficiently to allow the wax to be melted out, and the shell after heating (baking) hardens further into a ceramic mould. In either case, the nature of the moulding mixtures and the method of mould construction impart specific characteristics to such moulds which differ from those of more conventional sand moulding. For example, these methods of mould making are not readily adaptable for heavier castings, but on the other hand the dimensional accuracy of the mould cavity

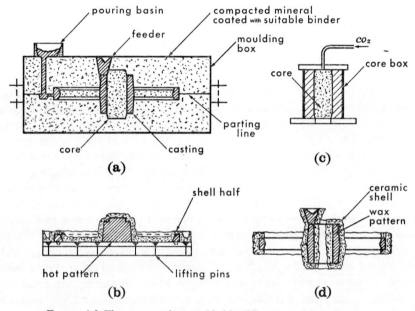

FIGURE 1.3 The same casting moulded in different moulding mixtures.

(a) Sand mould and clay binder.
(b) Shell mould: sand base and synthetic resin binder; two halves joined give mould identical to (a).
(c) Core for mould (a) or (b): sand base, sodium silicate binder hardened with CO_2.
(d) Ceramic shell mould: sand base, silica binder.

and of the casting produced is generally greater than that obtained with conventional sand castings.

(c) *Metal (permanent) moulds or dies.* Instead of obtaining the mould cavity by means of a pattern and a mineral moulding mixture, a metal mould or die is obtained by machining out in metal blocks (e.g., castings or forgings) the contour surfaces of the article required in such a way that such blocks assembled together provide the required mould cavity. The liquid metal may be forced into such moulds at high pressure (to ensure mould filling and enhance dimensional accuracy of the casting), this process being known as pressure die casting, Fig. 1.4*a*; alternatively it may be poured under gravity, this being called gravity die casting, Fig. 1.4*b*. Metal moulds characteristically impose their own advantages and limitations in comparison with other moulding processes. For example, metal moulds are restricted for making castings in relatively low melting point alloys (up to approx. 1,200°C), whilst

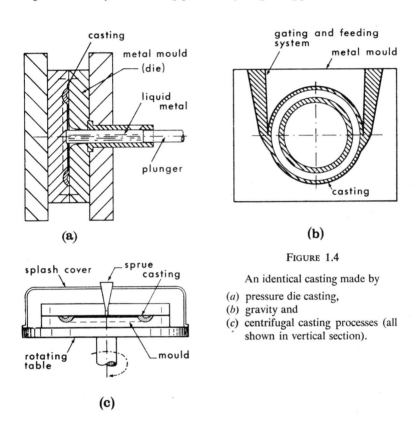

FIGURE 1.4

An identical casting made by

(*a*) pressure die casting,
(*b*) gravity and
(*c*) centrifugal casting processes (all shown in vertical section).

the withdrawal of metallic cores defining internal surfaces of castings imposes further restrictions on the choice of casting design. As a rule metal moulds are used for relatively simple casting shapes and for small weight castings. The main advantages of the process lie in the higher dimensional accuracy of the castings, higher productivity and improvement in some aspects of the metallurgical properties.

(*d*) *Centrifugal casting.* When centrifugal force, instead of the more conventional gravity or mechanical pressure force, is applied for mould filling with metal, ceramic or sand moulds, the process is referred to as centrifugal casting, Fig. 1.4c. It is used primarily when this specific method of mould filling is advantageous (for example, in casting pipes or similar simple geometrical shapes or very small articles). However, when improved metallurgical properties result from the application of centrifugal force during the solidification, the process is also used for making shaped castings. This process is much more restricted in scope in comparison with the other methods, and its general field of application is therefore relatively limited.

(e) *Developments in and central problems of moulding.* Although certain mould-making processes have been practised for centuries, they are continually being developed, and improvements in the old methods or even more radical innovations are still taking place. For example, shell moulding is barely a quarter of a century old, and some processes like that of using 'fluid' sand (Chapter 8) are of very recent origin. The opportunities for further developments in moulding, and indeed in other branches of founding, are as high as in any other field of metallurgy.

The central problems in the application of moulding and casting methods are:

(1) Correct selection of the most appropriate moulding and casting method for the casting designs where any one out of several methods is feasible.

(2) Utilisation of the selected method to obtain optimum properties of castings and at an optimum economy.

The object of moulding is to produce moulds of the required properties at a suitable cost. The properties and costs not only differ from one process or method to another, but even more frequently depend on the degree of skill and understanding in achieving the best results with any one method. A given casting design can be moulded and the mould constructed in a number of different ways, and this in itself presents a challenge to human ingenuity in arriving at the optimum solution.

The background to the knowledge necessary to find the answers to such fundamental problems of founding is considered in the next section.

1.4 FOUNDING: CRAFT, ENGINEERING AND APPLIED SCIENCE

Most of the technical definitions encountered in founding are straightforward and simple. It is only when one attempts a quantitative or phenomenological explanation of various problems that the more complex nature of the basic knowledge of founding becomes apparent. For a textbook presentation of this knowledge it is important to examine its origin and its background. An explanation of the basis on which the selection and grouping of this knowledge can be made is also desirable.

As founding is essentially a human activity, the knowledge or skill of 'how to make' castings can be gained by practical training. On the other hand, the knowledge necessary to understand 'how castings are or can be made' is obtained by a study of the process in terms of the engineering or scientific concepts involved. For educational purposes, one can learn how to make things first, and this can be followed by a study of the basic knowledge. Equally, the study may precede or be accompanied by practical training. In the foundry industry today the number of people employed as skilled craftsmen is relatively small, important and valuable though they are, whilst the number of people within and outside the industry concerned with the 'know why' of founding is very much larger. In general, the number of people directly and primarily involved in the 'know how' or skill aspects of producing castings is decreasing, whilst the number of people indirectly involved in the technology of the production, use and application of castings is continually increasing. The practical knowledge of founding can be generally summarised in terms of empirical rules obtained by practising the craft. The understanding of the various problems arising in the practice of this craft is, on the other hand, to be sought in the basic principles of engineering and metallurgical science. While different approaches to founding may proceed along different routes, and indeed may have different objectives, a full knowledge of the whole subject must rest on an understanding of the cardinal points of all these different approaches.

1.4.1 Foundry craft

Knowledge of founding began and grew with the practice of founding as an empirical craft. Recorded and unrecorded experience of how best to make moulds and cast metals has been passed from one generation to another and continuously improved upon. Today this knowledge is an essential element of foundry practice. Foundry craft, like any other craft, is mastered by practice, although certain aspects of it would be appreciated and understood by reading a suitable text. One of the major limitations of the craft approach alone to the knowledge of founding is that a single span of life is too short for anyone to become a complete foundry craftsman in all the different branches of modern founding.

1.4.2 Foundry engineering

As soon as one accepts that one of the objectives of founding is that of producing a large quantity of castings at an optimum economy, or that of meeting more exact requirements in the properties of cast metals, it becomes necessary to discuss founding in terms and concepts of engineering. Examples of such engineering problems extend from those of the design and performance of machines and plant used in the processes to those of performance and properties of castings. Because of this importance of engineering in relation

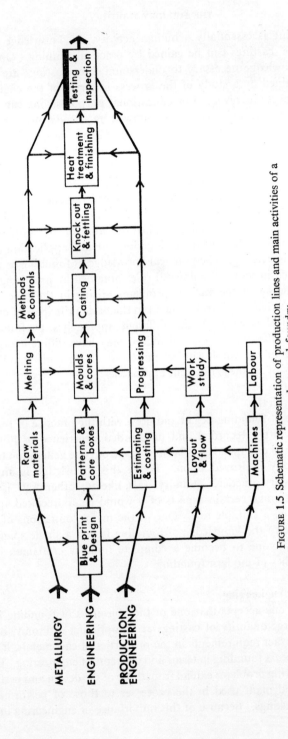

FIGURE 1.5 Schematic representation of production lines and main activities of a modern sand foundry.

METALLURGY

ENGINEERING

PRODUCTION ENGINEERING

Blue print & Design

Raw materials

Melting

Methods & controls

Patterns & core boxes

Moulds & cores

Casting

Knock out & fettling

Heat treatment & finishing

Testing & inspection

Estimating & costing

Progressing

Layout & flow

Work study

Machines

Labour

to skilled foundry practice, the foundry engineer, like the foundry craftsman, is an essential human ingredient in the structure of the industry.

A study of founding can therefore be made independently in terms of engineering concepts, so long as one accepts clearly at the start that such an approach deals with only a part and not with the total field or knowledge of founding.

1.4.3 Applied science of founding

Founding can be looked upon as a process of converting some metallic raw materials into useful products. This in turn calls for the use of other materials at various stages of the process. Many of the problems that are encountered can be best understood and analysed in terms and concepts of applied science, mainly those of branches of physics and chemistry. For example, in a foundry process metals are heated and cooled in different environments. Various chemical reactions which can occur can be treated by applications of the laws of physical chemistry. Similarly, the formation of cast structure from the liquid during the solidification of a casting can be understood and controlled by applying the theories of formation and growth of crystals, which are branches of physical metallurgy. Heating and cooling of moulds too brings in examples of physical or chemical problems. However, as with the craft and engineering studies of founding, the applied science approach alone embraces a part and not the total knowledge of founding.

Most of the essential activities of founding can be summarised in the form of a chart, such as that shown in Fig. 1.5. In the practice of founding at the present time, some of the activities shown in the chart can be adequately and best executed by the application of empirical knowledge; for example pattern making or mould construction. On the other hand, problems of the design of melting furnaces or machines for mould making, or of the sand plant for moulding mixture preparation, call for the application of engineering knowledge. In the same way, problems of metal control and behaviour at various stages of founding require the application of knowledge of the science of metals and other materials. The interconnections of the various activities in founding are indicated by arrows in the chart. This serves to emphasise the close interdependence of the craft, engineering and metallurgical approaches to the modern practice of founding.

FURTHER READING

Historical

AITCHISON, L., *A History of Metals*, Macdonald & Evans, London, 1960.

SMITH, C. S., *A History of Metallography*, University of Chicago Press, 1960.

COGHLAN, H. H., *Prehistoric Metallurgy of Copper and Bronze*, Oxford University Press, 1955.

General textbooks of founding

HEINE, R. W. and ROSENTHAL, P. C., *Principles of Metal Casting*, McGraw-Hill, New York and Maidenhead, 1955.

ROLL, F., *Handbuch der Giesserei Technik*, Vols. I and II, Springer-Verlag, Berlin, 1959 and 1963.

TAYLOR, H. F., FLEMINGS, M. C. and WULFF, J., *Foundry Engineering*, Wiley, New York and London, 1959.

FLINN, R. A., *Fundamentals of Metal Casting*, Addison-Wesley, Reading (Mass.) and London, 1963.

HOWARD, E. D., *Modern Foundry Practice*, Odhams Press, London, 1958.

MORRIS, J. L., *Metal Castings*, Prentice-Hall, Englewood Cliffs, N. J., 1957.

COOK, G. J., *Engineering Castings*, McGraw-Hill, New York and Maidenhead, 1961.

AITCHISON, L. and KONDIC, V., *The Casting of Non-Ferrous Ingots*, Macdonald & Evans, London, 1953.

BOICHENKO, M. C., *Continuous Casting of Steel*, Butterworth, London, 1961.

CHAPTER 2

MELTING OF METALS

2.1 INTRODUCTION

The main steps in a typical casting process are illustrated in Fig. 2.1. A charge containing solid pieces of metals and alloys of various shapes, sizes and chemical purity, is melted in a refractory lined furnace by supplying the

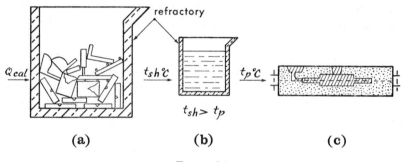

(a)   **(b)**   **(c)**

FIGURE 2.1

(*a*) Furnace containing ingot or scrap charge heated to the molten metal temperature (superheat), t_{sh} °C.

(*b*) Ladle filled at intervals and emptied with the metal at the pouring temperature, t_p°C.

(*c*) Mould.

required quantity of heat. Molten alloy at a suitable superheating or tapping temperature, t_{sh}°C, is transferred to a ladle, from which moulds are filled at the pouring temperature, t_p°C. When convenient, moulds can be filled directly from the furnace.

2.2 COMPOSITIONAL PROBLEMS

The essential metallurgical problem encountered in the melting process is that of the changes in metal composition. In order to produce a casting with the required properties, the alloy melted must be of a specified composition. However, the final alloy composition is normally obtained from a charge of a different initial composition which alters in the course of the melting process. Furthermore, the molten alloy must have, prior to pouring, not only the specified elements within the required limits, but also the unwanted or undesirable elements—impurity content—below the specified limits.

With very few exceptions metals exist in nature as simple or complex chemical compounds, often combined with oxygen or sulphur. Extraction and refining metallurgy, which are based on the application of the laws of chemical thermodynamics, deal with the problem of obtaining 'pure' metals from such compounds. The term 'pure' is a relative one and is defined in 2.4.2. These laws also explain why metals return to such compounds whenever the external conditions allow this to occur. Such conditions are encountered in most metal melting processes, which should therefore be carried out in such a way as to prevent or reduce the extent of the melt combining with the elements that surround it. These are derived from the following:

(*a*) The *normal atmosphere* (O_2, N_2, H_2O and CO_2), as for example in electric furnace melting, where no special protection of the melt is applied.

(*b*) The *combustion products* of oil, gas or solid fuels used for melting (CO_2, CO, H_2O, SO_2, with O_2 and N_2 from the air); with solid fuels the elements found in the ashes, such as phosphorus and sulphur, may also act as a contaminating source.

(*c*) The *refractory containers* in which the metal is melted, if they are not completely inert to the melt; metallic and non-metallic contamination with, for example, silicon, aluminium, oxygen or hydrogen can result from this source.

(*d*) *Materials in or on the charge* and *various tools used as melting aids*; scrap, for instance, may be covered with sand, rust or corrosion products; in the form of swarf it may also contain other metallic or non-metallic elements which cannot readily be separated; melt stirring, degassing and sample-taking tools are also frequently used.

In any specific melting process, therefore, the final composition depends upon the composition of the charge together with any changes during melting.

From the point of view of compositional changes, melting processes fall into two major groups: simple or straight melting processes, and melting-refining processes. In the former group are those melting processes where relatively small compositional changes occur during melting, and very little, if any, adjustment is needed prior to pouring. Typical examples are air melting of some low melting point alloys and light alloys, or vacuum or protective atmosphere melting of high melting point alloys. In a process which combines melting and refining, the molten charge has to be adjusted in order to obtain the required composition. For instance, the carbon content of steel, and often the level of other elements as well, is lowered during melting by oxidation and oxidising slag reactions; the oxygen content of the melt is subsequently corrected by the use of deoxidants.

The wide range of raw materials, fuels and furnaces has led to the development of a large number of melting processes now in use for different alloys. The point has already been made that both technical and economic factors must be weighed when selecting a particular founding process. To illustrate this in relation to melting, titanium is chemically highly reactive and therefore requires a process that completely excludes the normal atmosphere and uses an inert gas or a vacuum. On the other hand, the chemistry of cast iron is less restrictive, so that here the more relevant question is that of choosing the most economical melting process that will give castings with the required properties. The same applies to the various melt treatments, such as fluxing, slagging, degassing, deoxidation and grain refining (2.4).

2.3 METALLURGICAL PRINCIPLES OF MELTING

Compositional changes occurring in the melting process as a result of chemical reactions can be analysed by applying the laws of homogeneous and heterogeneous equilibria and of reaction kinetics, which are subjects of chemical thermodynamics. Other changes can be explained more directly in terms of pressure and temperature variables, i.e., of the physical changes and behaviour of metallic liquids.

2.3.1 Chemical reactivity

A solid metal A is contained in a refractory crucible. During heating to its melting point the gases in contact with the metal may react with it, and this reaction is particularly likely to occur with oxygen. The reaction between solid metal and solid refractory is so slow in most cases that it can be neglected. The solid or liquid metal reaction with oxygen can be represented by a chemical equation of the type

$$M + O_2 \rightleftarrows MO_2$$

The oxygen may either dissolve in the liquid metal or collect as a film or powder on the surface which later may enter a slag or flux (2.4.2). Dissolved oxygen is usually detrimental. Oxygen control and removal is effected by special techniques such as vacuum melting and deoxidation treatment

(Chapter 4). The oxidation reaction, as already indicated, may be used with some alloys to remove from the melt unwanted elements such as carbon, sulphur and phosphorus, which oxidise preferentially to iron; this is done in oxygen refining of steel.

The readiness with which one element will react with another can be expressed in terms of quantities defined as chemical potentials. In every reaction an energy change is involved; the chemical potential is a measure of the energy which is available or free, and can be considered as the driving force for the reaction. Once the potential or free energy of reaction of a number of elements has been determined, their mutual chemical reactivity

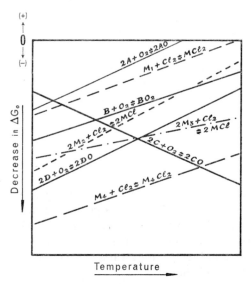

FIGURE 2.2 Standard free energy—temperature diagram.

can be compared. Comparative data can be presented by means of free energy or chemical potential diagrams (Fig. 2.2).

A reaction occurs when there is an overall decrease in the free energy of the system, or in other words when the total free energy of the products, ΔG_{prod}, is smaller than that of the reactants, ΔG_{react}. The change in free energy is defined as $\Delta G = \Delta G_{prod} - \Delta G_{react}$, and the value of ΔG of a spontaneous reaction is therefore negative. The greater the change in the free energy of the system, the greater is the driving force of the reaction. This is illustrated in Fig. 2.2.

The diagram shows that the magnitude of the free energy change is also temperature dependent. For metals A, B and D the effect of an increase in temperature is to promote a decomposition of their oxides. This means also that at high temperatures the oxides of the refractory metals which are used as melting containers could be appreciably less stable and could become a source of melt contamination. In examples of this kind the effects of pressure and other controlling factors of reaction kinetics have also to be considered.

The free energy diagram example shows that, with an alloy of the two metals A and B, metal B will tend to oxidise before A. Similarly, if metal B is added to molten metal A containing dissolved oxygen, then B will tend to take most of the oxygen available, and if the oxide BO_2 is solid at this temperature it will tend to separate from the melt. This is the principle applied in the deoxidation of many alloys.

Free energy diagrams are available for many chemical reaction systems; for others the free energy change can be obtained by calculation. These provide the main sources for a qualitative understanding and numerical estimation of the outcome of the chemical reactions that occur in metal melting processes.

Free energy relationships are postulated on a somewhat idealised behaviour of elements in chemical reactions. In most alloy problems such ideal behaviour does not obtain and the degree of departure from ideality has to be established experimentally. This limits the practical usefulness of free energy methods for accurately calculating composition changes in metal melting. The magnitude of the free energy change indicates the driving force of a reaction, but it does not follow that it will proceed to or attain completion unless the kinetic or speed-controlling factors are also favourable. For example, both aluminium and magnesium readily form oxides. With aluminium a protective film of Al_2O_3 on top of the melt virtually stops the reaction, and aluminium can be melted in air without special precautions. No such continuous film forms on liquid magnesium, which consequently has to be covered with a flux or kept under a special atmosphere to prevent complete oxidation.

2.3.2 Mixing and equilibrium

One of the objects of the melting process is frequently that of mixing two or more metals into an alloy. One or more alloying elements, known as solutes, are dissolved in a parent metal, which is called the base or solvent. There is no simple physical or chemical property that can be used to predict the degree of solubility of one metal in another, but equilibrium diagrams such as Fig. 3.5 (page 42) provide conveniently the necessary information. The figure shows that, if the two metals are completely soluble in one another in the liquid state, one liquid phase is indicated over the whole range of composition of metals A and B. If the metals only partially dissolve in one another, then two liquid phases will be shown. Binary and ternary diagrams are available for most metals and also for some metal–non-metal systems, for example metal–oxygen. The solubility of gases which dissolve in both liquid and solid metals, such as hydrogen in aluminium or iron, is conveniently shown by the solubility diagram (Chapter 4). An equilibrium diagram also indicates the temperature at which a liquid alloy starts to freeze or solidify. This is important for determining the superheat temperature required to carry out the casting process satisfactorily, which is usually of the order of 100°C to 300°C above the liquidus. Further important uses of the equilibrium diagram are discussed in Chapter 3.

2.3.3 Physical properties

Vapour pressure. Molten metals, like all other liquids, have their own vapour in equilibrium with the liquid. The pressure of this vapour varies with the temperature and, in alloys, with the composition. A few metals, such as cadmium, zinc, magnesium and antimony, have such a high vapour pressure at the melting temperatures of required alloys that with an external pressure of 1 atm special precautions have to be taken when they are used as solvents or solutes. At temperatures close to the melting point the vapour pressure is still fairly low, but with increasing temperatures considerable losses may occur through evaporation. At these temperatures special melting techniques have to be used (2.4.4).

Melting temperature. Melting temperatures of metals and alloys commonly used in industry vary from 200°C to 1,600°C. When a solid metal A with a low melting point is added to a liquid metal B with a high melting point, if A is soluble in B, then provided the vapour pressure of A is not too high no special mixing problem arises, although mechanical stirring to achieve homogeneity may be needed. But if liquid metal A is the solvent and B is used as a solute, the high melting point of B will give rise to considerable difficulty in alloying. This can be overcome by first making, in a special furnace, a richer alloy of B in A, which may consist of 50% of each by weight. The resulting temper alloy or hardener is then used more readily for making smaller additions of B to A in normal foundry processes; for example, this is the method employed for adding heavy metals such as iron, nickel or manganese to aluminium base alloys. Temper alloys are often used to simplify compositional control of alloys in production.

Heat content. The heat content of molten metal at a given temperature is made up of the heat required to bring the metal to the melting point, the latent heat of fusion and the heat required for superheating. These have been determined experimentally for some metals and are given in the standard handbooks. Such data are used for calculating the heat requirement of melting furnaces, and for solving problems of heat flow in the cooling of castings, but generally they are available at a high standard of accuracy only for pure metals and for a few alloys; for others approximate values are used. Heat of solution (positive or negative) may to a marginal extent alter the temperature of a melt on alloying. A large contribution to the temperature changes of the melt results in some refining processes where the oxidation of impurities is accompanied by substantial generation of heat, this being essential in some processes, e.g., Bessemer steel making.

The metallurgical principles briefly summarised in the preceding paragraphs indicate that a knowledge of thermodynamics is essential for understanding fully the problems encountered in metal melting. Approximate calculations of compositional changes can be made assuming ideal behaviour, but a quantitative treatment of non-ideal systems is only possible once the degree of departure from ideality has been established experimentally. Before considering the application of these general principles to practical problems associated with changes occurring in a metal during melting, it is necessary to examine certain essential engineering and production characteristics of melting processes.

2.4 MELTING PROCESSES

2.4.1 Melting furnaces

A foundry melting furnace can be defined as a melting unit which has to bring a given weight of metal of required or specified composition to the tapping temperature at the required melting rate and economic efficiency. These are the five factors from which the optimum melting process for a foundry may be determined. Apart from the problem of alloy composition, which is discussed in the present chapter, the remaining problems of melting furnaces are discussed in the further reading. However, a brief outline of melting furnaces is given here to bring out some factors pertinent to the composition problem. A broad summary of the types of furnace used in industry, showing typical temperature, capacity and melting rates, is given in Figs. 2.3 and 2.4.

From the point of view of basic design and of metal tonnage, the two main groups are batch and continuous furnaces, whose characteristics are shown in Fig. 2.3. The melting weights and rates indicated are for ferrous metals. Weight capacities for non-ferrous metals can readily be obtained by multiplying by the density ratio ρ_{nf}/ρ_{fe}. The capacity of melting furnaces is to some extent determined by heat energy conversion factors, but for many furnaces the upper limit is largely controlled by production considerations.

The working temperature range of a furnace depends on its design and the method of heating used. Electrical heating methods vary according to the temperature requirements, Fig. 2.4. This type of furnace gives the highest degree of control of the furnace atmosphere, freedom in the choice of the refractories and therefore in the control of the melt composition itself, and is suitable for straight melting of alloys sensitive to compositional change. Furnaces fired by combustible fuels are more generally used for refining–melting processes (2.4.4) and for those melting processes where the compositional problem can be adequately controlled.

The melting rate of a furnace is determined by the furnace capacity, type of charge, heat transfer conditions and melting temperatures required. From Figs. 2.3 and 2.4 it can be seen that fuel-fired crucible furnaces melt more slowly than hearth furnaces, and that resistance furnaces are slower than induction or arc furnaces.

The economic efficiency of melting cannot properly be isolated from the economy of a foundry process as a whole. Important factors are running

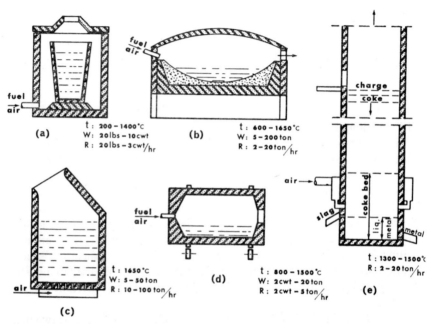

FIGURE 2.3 Typical fuel-fired furnaces showing working temperature range *t*, capacity *W* and melting rate *R*.

Batch types: (*a*) crucible, (*b*) hearth (reverberatory, incandescent), (*c*) converter (for refining premelted irons into steel by oxidation), and (*d*) rotary (reverberatory).

Continuous type: (*e*) cupola.

costs (fuel, refractories, maintenance and labour), metal melting losses and capital depreciation.

2.4.2 Basic elements of melting

Charge. The charge may consist of one or more of the following ingredients: pure metals, pre-made temper alloys, pre-made alloys of specified composition (ingotted alloys) and previously used metal (scrap). It is possible to purify some metals by refining techniques such as zone refining (see Chapter 6) and obtain metals more than 99·9999% pure or having less than

1 p.p.m. (one part per million) of impurities. In industrial practice metal purity is usually far less than 99·9999%, and the terms 'pure metals', 'secondary alloys', 'refined ingots' and many others are used to denote metals of a specified purity or origin. The purity of most industrially used metals varies from 99·5% to 99·9%, but there are exceptions at both ends of the scale. The selection of the charge is made on the principle that castings must meet compositional specification standards (Chapter 11). As the cost of

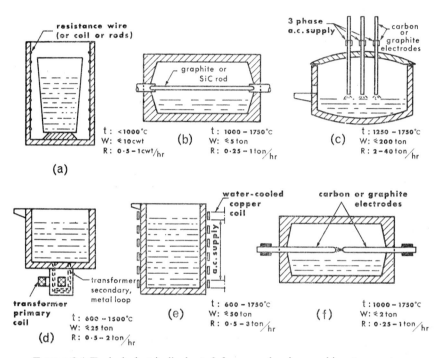

FIGURE 2.4 Typical electrically heated furnaces showing working temperature range t, capacity W and melting rate R.

(a) resistance, (b) rod resistor, (c) direct arc, (d) low frequency channel, (e) high or low frequency crucible, and (f) indirect arc.

metals increases rapidly with their purity, it is clear that the alloy specifications and metal costs jointly determine the choice of raw materials for the charge in making a given alloy. The metallurgical problem is to control the composition during melting and obtain optimum properties from metals of a selected purity. With low melting point and simple binary alloys, charges are usually made up of pure metals, temper alloys and controlled scrap. In contrast to this, steel melting practice is largely based on refining scrap charges, which may however include some temper alloys as well as ingotted alloys in the form of pig iron. For many complex alloys containing several elements, such as many light alloys, bronzes and special steels, it is more economical to pre-make ingots to the correct and full specification analyses and to carry out only a simplified remelting process in the foundry. This is called ingotted alloy melting practice.

Refractories. Refractory linings of furnaces and ladles used for molten metals should satisfy the combined requirements of long life at the operating temperatures and of chemical inertness to the melt and any slags or fluxes that are used. Refractories are chemical compounds or mixtures, usually of oxides, silicates, aluminates and carbides, which are non-reactive or resistant to liquid metal, slag or flux solutions at high temperatures. The subject of refractories is too wide even for a brief summary in the present text, and further reading in this field is strongly recommended to all students of founding.

Slags and fluxes. The various sources of contamination of the melt have already been indicated. Those impurities and reaction products which are not soluble in the melt form another liquid layer at the surface known as slag. Typical slag, consisting mainly of oxides and silicates and a minor content of other compounds, is particularly associated with melting or refining of ferrous metals. With non-ferrous metals a similar liquid or solid layer is frequently formed intentionally by fluxing, i.e. by the addition of suitable ingredients (fluxes), to dissolve solid metal oxides (dross) and other impurities at the liquid metal surface. In addition, slags and fluxes perform useful functions by protecting the melt from the surrounding atmosphere, or by assisting chemical reactions; this means that they are widely used in refining for removing unwanted elements or for introducing into an alloy certain solutes that present difficulties of alloying by other methods. The terminological distinction between slags and fluxes is somewhat blurred, since some specific additions often have to be made to give a slag of the required characteristics, and from a metallurgical point of view slags and fluxes have virtually the same functions. Both slags and fluxes contain the unwanted products of melting with such extra additives needed to obtain the desired physical and chemical properties of the liquid layer (melting point, fluidity, density, solubility and chemical reactivity).

Where slags or fluxes have a strong basic or acid character special refractories are needed to avoid rapid chemical erosion of the furnace walls (basic or acid melting practice of alloys).

2.4.3 Stages of melting

Melting sequences differ in detail from one alloy to another, but the general principles can be summarised as follows:

(*a*) *Melting down.* As the temperature is raised, the surface of the charge may start reacting with the surrounding atmosphere. Although solid-gas compared with liquid-gas reactions are slow, the time of melting is usually sufficiently long, particularly in ferrous practice, for compositional changes to occur during the melting down. This is particularly so when melting scrap having a large surface area, e.g., swarf. Even with a briquetted swarf the amount of oxide dross produced necessitates, with most alloys, some fluxing or slagging treatment.

(b) *Melt treatments.* When all the charge is melted down, for some alloys and melting conditions the only further step is to superheat the metal to the pouring temperature and to pour the casting. With many industrial alloys, however, fluxing and slagging operations are necessary for the removal of impurities and undesirable metallic or non-metallic elements dissolved in the metal or floating at the liquid metal surface.

(c) *Compositional adjustment.* When physical factors demand it, final alloy additions to the melt are made towards the end of the process. For example, final adjustment of magnesium content in aluminium alloys, or of phosphorus in some bronzes, is best made at a predetermined time and temperature before pouring the alloys, in order to avoid losses in these elements. Similar adjustments may also be necessary in refining-melting of steel. At this stage it may be important to know the actual composition of the melt as determined by one of the methods discussed in Chapter 11.

(d) *Properties adjustment and control.* As well as showing whether the alloy will meet the standard specifications, the melt analysis or special tests may reveal impurities, particularly of a gaseous nature, which may affect the properties of the casting. Adjustment of properties can be effected by deoxidation and degassing (Chapter 4). Grain refining and inoculation treatments may also be employed to control the crystallisation behaviour during solidification (Chapter 6). Most of these treatments are carried out just prior to pouring, otherwise their useful effect may be lost.

2.4.4 Classes of melting

Vacuum and inert atmosphere melting. The main object of vacuum melting is to reduce the extent of chemical reactions of the melt with the gaseous atmosphere surrounding it. Vacuum pressures of 10^{-2} to 10^{-4} mmHg are generally used, and therefore the process can only be applied to metals and alloys of very low vapour pressure. When it is necessary to prevent loss by evaporation of some constituents, the melting chamber may be brought up to atmospheric pressure by refilling with a chemically inert or neutral gas, such as argon, helium or nitrogen. These are the optimum melting processes for reducing contamination of the charge, but to obtain the best results special care is needed in the selection of suitable refractories. An important use is for producing steels and similar alloys for high-temperature and high-stress applications, where the presence of impurities derived from a normal atmosphere would have a serious effect on the structural quality of the products.

Air or straight melting. For some metals and alloys the contact between air or combustion gases and the melt for a limited time is not serious. Many tin, lead, zinc, aluminium, copper, nickel and iron alloys are in this group. The extent of reaction with oxygen or water vapour in some cases may necessitate partial surface cleaning with slags or fluxes. Where oxygen or hydrogen are soluble in the alloy, either deoxidation or degassing, or both,may be necessary.

Refining-melting. Large tonnages of alloys are produced from scrap metals. Non-ferrous scrap is usually refined by specialised processes, and ingotted alloys of required composition are supplied to the foundry industry. Ferrous practice is based more frequently on refining the scrap directly in the foundries by refining-melting. The melting process is based on mixed charges (pig iron and scrap of various kinds), and compositional changes in melting may be appreciable. It is typically applied to grey irons, carbon steels and some alloy steels. Grey iron is a borderline case between straight and refining-melting; the problem is to obtain the right composition with the least amount of refining, with special reference to variations in carbon, silicon, manganese, sulphur and phosphorus contents. For ordinary carbon steels the charge is fully refined; all or most of the alloying and tramp elements are reduced by oxidation to the desired low level, and then the composition is readjusted by special additions.

It is clear that some compositional changes are inevitable in all melting processes, these being smallest with vacuum melting and greatest with refining-melting. If other factors permit a free choice of melting method, this will depend on the tolerances in the composition of the charge and of the final product, and the degree of change in composition which may have to be controlled during melting.

2.5 APPLICATION OF METALLURGICAL PRINCIPLES

The physical and chemical principles discussed can be applied to a number of problems arising in melt composition controls. These include the selection of raw materials for charges and refractories, the control of furnace atmospheres, and the control of the effects of process variables such as pressure, temperature and concentration. At the present time the essential nature of melting processes is well understood, but the extent to which quantitative calculations can be applied in practice varies a great deal.

2.5.1 Chemical reactions

Figure 2.5 shows how the laws of chemical equilibria can be applied to the three main classes of melting.

Referring to Fig. 2.5a,

(i) It is possible to calculate the partial pressure of oxygen above the melt that can be allowed without oxidation of the molten metal; this should be smaller than the dissociation pressure of the oxide of the metal in question.

(ii) It is possible to foresee whether the molten alloy will tend to reduce the oxide of its refractory container, and what will be the effect of solute concentration on the direction and magnitude of the reaction.

(iii) Equilibria calculations can be used to show whether the melt can be purified by vacuum melting; vacuum degassing is discussed in Chapter 4.

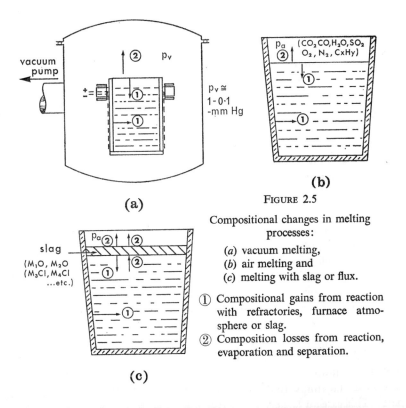

(a)

(b)

FIGURE 2.5

Compositional changes in melting processes:

(a) vacuum melting,
(b) air melting and
(c) melting with slag or flux.

① Compositional gains from reaction with refractories, furnace atmosphere or slag.
② Composition losses from reaction, evaporation and separation.

(c)

Fully quantitative calculations of compositional changes for air or furnace atmosphere (straight) melting processes are extremely difficult, because of the complex nature of the systems and the lack of free energy data. The nature of the problem is indicated in Fig. 2.5b. Some of the gases present may react with the molten metal, others may be neutral; the standard handbooks give some of the necessary free energy data. The tendency for oxygen, hydrogen, sulphur, nitrogen and carbon to enter the molten metal can be predicted from appropriate graphs or by calculation, although it is generally impossible to forecast the actual content of these elements in the molten metal because of the departure from ideal behaviour and because of the uncertainty of the kinetic factors controlling the time required to reach the equilibrium.

Figure 2.5c illustrates the situation where a flux or slag is used. Here too the basic information is available from free energy diagrams and tables. Figure 2.2 gives some hypothetical free energy data for chlorides. The chlorides of M_1 and M_2 appear in the diagram above those of M_3 and M_4, indicating the greater stability of the latter. If higher placed chlorides are used to cover a molten alloy containing metals M_3 and M_4 a contamination of the melt will result. Conversely, chlorides of M_3 and M_4 can be used to cover a melt containing M_1 and M_2, since the latter will not appreciably reduce the more stable chlorides. However, since these reactions are reversible, some slight contamination may occur.

The following illustrates these principles. Cover fluxes for aluminium base alloys are made up largely from sodium and potassium chlorides and fluorides. As long as chlorides predominate in the flux mixture, there is little contamination of the melt with sodium or potassium, but when the fluoride content is increased some sodium may enter the alloy. This may be beneficial, as in grain refinement of silicon in aluminium–silicon alloys (Chapter 6), or harmful as in embrittlement of aluminium–magnesium alloys. Another example is the alloying of manganese or titanium with aluminium by plunging manganese chloride or potassium or sodium titano-fluorides into the melt (these elements can also be added as pure metals or as temper alloys, but not so conveniently if required in small amounts).

It must be emphasised that other factors besides equilibria are significant in relation to the selection of slags and fluxes, namely melting points, vapour pressures and densities, and the toxicity and corrosive behaviour of the various salts and oxides used. Furthermore, their beneficial effects have to be balanced against the extra melting cost. In practice, since they have to meet many conflicting requirements, the selection of suitable chemical compounds for fluxing is limited to relatively few chemical compounds.

It might be thought that the importance of chemical equilibria data in refining processes would be very much greater than in straight melting, and yet it is exactly here that theory and practice are at present most widely apart. This is due largely to the complex nature of the systems involved; molten iron, for example, contains carbon, silicon, sulphur, phosphorus and manganese, while the slag contains silicon, aluminium, calcium, iron, oxygen, sulphur and phosphorus, largely as solutions of various oxides. The behaviour of such a system has to be determined experimentally. This is difficult, as it frequently involves temperatures of over 1,500°C. At present, therefore, some of the features of complex refining systems are not fully understood, nor are all the necessary calculation data available. But, in general, the behaviour of such systems can be qualitatively understood and approximate calculations can be made. For example, it can be shown that sulphur removal from molten iron is dependent on the oxygen potential in the slag and the metal and requires a low oxygen content in the iron. Thus the scientific study of practices that were developed empirically has frequently made it possible to introduce modifications in refining processes resulting in considerable production improvements.

2.5.2 Physical data

In comparison with the complexity of some chemical reactions encountered in melting processes, the physical problems are relatively straightforward. If an alloy containing elements X, Y and Z is to be melted, and if X has a high melting point and Z a low melting point but a high vapour pressure, it may be impracticable to mix the three elements directly. The solution may lie in premixing the most favourable compositions to produce temper alloys, or in adding the volatile element (under special precautions) just prior to pouring the molten metal. In some extreme instances, the only practical answer is to prepare such an alloy in the solid state, using powder metallurgy techniques.

When the vapour pressure in melting is not very high, a straight addition of the more volatile metal can be made. For example, in making 60/40 brass, zinc losses due to volatilisation can be kept below 0·3% by avoiding unduly high melt temperatures and by using rapid mechanical stirring of the zinc in the molten copper; whereas, when magnesium is added to molten iron during the manufacture of nodular irons, the vapour pressure of magnesium is so high that either a sealed-off ladle is used or temper alloys containing 10% to 20% magnesium are applied. Similarly, temper alloys are used to overcome a vapour pressure problem in preparing some copper alloys containing phosphorus; a temper alloy containing 5% to 15% phosphorus is extensively used for alloying or for deoxidation of these copper-base alloys.

Apart from the physical hazards and difficulties, there is a danger of the metal vapour being toxic, beryllium being an extreme example.

Melting is normally the most convenient way of obtaining a homogeneous alloy in which all the added elements are uniformly distributed or dissolved in the solvent. However, when there are big differences in specific gravities or melting points, mechanical assistance such as stirring may be necessary to obtain homogeneity. When alloying elements are only partially soluble in the base for example lead in copper, production of a homogeneous alloy requires the use of special techniques at the liquid stage, followed by rapid cooling during solidification. The general problem of homogeneity is of great importance in the solid state and will be discussed further in Chapters 6 and 7.

2.6 SUMMARY

A study of the melting processes used in founding reveals that several problems are interrelated. For a general and optimum production solution it is necessary to consider metallurgical, engineering and production engineering factors. Most of the problems of chemical composition control can be resolved with the aid of basic sciences, but the solutions obtained have to be reconciled with the engineering factors of melting equipment performance and design and the production engineering operational factors (volume, rate of melting, superheating temperatures and costs). In planning a new foundry these problems can be considered by application of existing scientific and empirical data. In an established foundry, much can be gained by analysing the present practices in order to arrive at a general optimum melting solution.

FURTHER READING

MACKOWIAK, J., *Physical Chemistry for Metallurgists*, Allen & Unwin, London, 1964.
KUBASCHEWSKI, O. and EVANS, E. L., *Metallurgical Thermochemistry*, 2nd ed., Pergamon Press, Oxford, 1958.
DARKEN, L. S. and CURRY, R. W., *Physical Chemistry of Metals*, McGraw-Hill, New York and Maidenhead, 1963.
HUME-ROTHERY, W., *Atomic Theory for Students of Metallurgy*, Institute of Metals, London, 1947.
THRING, M. W., *The Science of Flames and Furnaces*, Chapman & Hall, London, 1952.
BUNSHAH, F. F., *Vacuum Metallurgy*, Reinhold, New York, 1958.
GILCHRIST, J. D., *Furnaces*, Pergamon Press, Oxford, 1963.
KINGERY, W. D., *Introduction to Ceramics*, John Wiley, New York and London, 1960.

CHAPTER 3

PHYSICAL AND CASTING PROPERTIES OF LIQUID METALS

3.1 INTRODUCTION

The physics of liquids is essentially concerned with the structure and properties of the liquid state in terms of binding forces and energies of atoms. Physical chemistry deals more generally with interactions of atoms of two or more different elements. Casting properties, on the other hand, are those that explain the behaviour of liquid metals in filling a mould, their subsequent solidification and their influence on the properties of the castings. It follows that they can be related to the physical and physico-chemical properties of metals, but they may also be affected by a sequence of other physical or chemical phenomena occurring during the casting process. In other words, the specific behaviour of a liquid metal found empirically to arise in a casting process can be scientifically related to and interpreted in terms of certain fundamental properties. Owing to the complexity of such properties their measurement and interpretation is in many cases empirical and qualitative. In order to discuss the casting properties of metals it is therefore essential to review briefly the physical and physico-chemical properties to which they are related.

3.2 PHYSICAL PROPERTIES OF LIQUID METALS

3.2.1 Structure

The structure of solid metals is crystalline, implying a regular fixed geometric arrangement of atoms held together by strong bonding forces (metallic and, with some metals and alloys, non-metallic bonds), Fig. 5.5. In the gaseous state the atoms are free to move about, and in a closed container their behaviour and properties can be analysed in terms of the kinetic theory of gases. In the liquid state the atoms are still mobile, but the bonding forces hold them together in an open container; the liquid state is, with respect to many of its properties, a transition between solid crystallinity on the one hand and gaseous, continuously changing randomness on the other. Some useful knowledge about the distribution of atoms in the liquid state can be obtained from X-ray reflection measurements. If a beam of X-rays of wavelengths λ is directed to a clean surface of liquid metal, at a varying incidence angle θ, the intensity of the reflection I can be plotted against the value of sin θ/λ, Fig. 3.1.

X-ray reflection data suggest that the distribution of continuously moving atoms in the liquid is such that their distance from a given momentarily fixed atom is not completely random. One of the proposed methods of

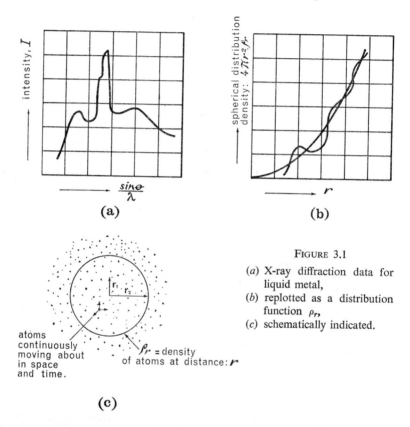

FIGURE 3.1

(*a*) X-ray diffraction data for liquid metal,

(*b*) replotted as a distribution function ρ_r,

(*c*) schematically indicated.

evaluating this distribution from the X-ray data is that of obtaining the radial density of distribution ρ_r (density per unit volume at distance r). This can be obtained by a mathematical analysis of $I - \sin \theta/\lambda$ graphs, and results in a plot of $4\pi r^2 \rho_r$ (spherical distribution), Fig. 3.1*b*. Such X-ray measurements show further that the interatomic distances in the liquid are not very much greater than those in the solid (this is in agreement with the relatively low values of volumetric expansion data of metals in melting, Table 3.1). Secondly, the density of distribution data suggest that some kind of transient spatial arrangement of atoms exists in liquids (the term 'quasi crystallinity' is often used), and several physical models for such arrangements have been proposed.

For the purpose of metal casting studies, the most interesting results of liquid structural experiments are those concerning the effects of temperature and of alloying. The X-ray data show that the density of distribution peaks gradually disappear with the rise in temperature, i.e. the liquids become generally more disorderly. On the other hand, the liquid structure—alloy constitution experiments reveal that in some pure metals (e.g., Ge or Sn) there exist two preferred distributions. It has been suggested that the second or extra distribution is due to the existence of clusters, i.e., regions in the liquid of a specific distribution due to differences in bonding, and hence an interatomic distribution different from that of the main liquid. With the addition of solute elements, certain anomalies have been shown in the X-ray reflection intensity graphs in liquids whose composition corresponds to some important phase change as indicated in the equilibrium diagram for the solid state (3.3.2 and 5.5). This again suggests some form of clustering, i.e., the existence of two different kinds of atomic distribution in the liquid. One kind of distribution differs from the other either because of differences in the nature of bonding, or because of the preferred distribution of atoms in regions where A–A, A–B or B–B bonds may predominate. For example, clusters could exist in the liquid of a eutectic composition, these corresponding to such a ratio of the two kinds of atom as occurs in the second phase of the eutectic in the solid state, as distinct from the ratio of the matrix. Structural studies of metallic liquids in general are of considerable importance for understanding the solidification behaviour of metals and alloys (Chapter 5).

3.2.2 Viscosity

A liquid metal will not flow if it is held in a horizontal tube which is closed at both ends by frictionless pistons. By using a small pressure on one of the pistons, the liquid will push the other piston and start to flow. The small force required to cause the liquid to flow is a measure of the weak bonds holding the atoms together in a liquid. The resistance against the flow is defined as viscosity or internal friction of liquids. The experiment of liquid flow in a tube can in fact be adapted to obtain numerical values of the viscosity of liquids. Such a method was used by Poiseuille for liquids at room temperature, and later developed by Sauerwald for low melting point liquid metals. In either case the liquid is made to flow in a fine capillary tube in such a way that the external force causing the flow is entirely used for overcoming the viscosity of the liquid. In Sauerwald's method (Fig. 3.2) the viscosity η is obtained by measuring the volume of metal flowing in a unit time and applying the relationship

$$\eta = \frac{\pi a^4 P}{8 Q l} - f(c)$$

where a is the radius of the capillary, P the mean pressure, Q the volume of liquid flowing in unit time, l the length of the capillary and $f(c)$ a kinetic energy correction factor. The unit of viscosity is called a poise (after Poiseuille) when its dimensions are expressed in the c.g.s. system.

The metallurgical problem of interaction between the walls of the capillary glass tube and liquid metals at a higher temperature, for example copper, nickel or iron, makes it necessary to use either an oscillating crucible or an oscillating pendulum method for viscosity measurement, Fig. 3.2. The method adopted is to measure the damping effect of the viscosity of the liquid

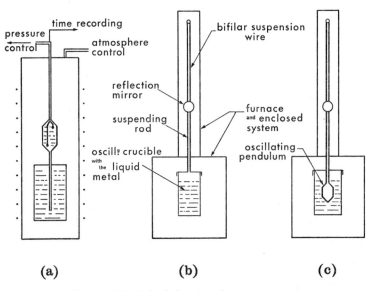

(a)   (b)   (c)

FIGURE 3.2 Methods for viscosity measurement.

(a) Capillary flow. (b) Oscillating crucible. (c) Oscillating pendulum.

on the oscillation of the crucible or pendulum. A mechanical oscillation is given to the system initially, and the damping is recorded by means of an optical system. As with the capillary method, the oscillation system of forces can be mathematically related to the viscosity of the liquid, or, what is equally frequently done, the experimental apparatus can be calibrated first by using liquids of known viscosity.

Owing to the experimental difficulties of the measurements (chemical inertness of materials used in contact with liquid metals), viscosities of some pure metals and only a few alloys are at present available. The main results of viscosity measurements of metallurgical interest are summarised in Fig. 3.3, which shows the variation of viscosity of a liquid metal with temperature and the changes of viscosity with composition of a typical binary alloy. The former relation was theoretically predicted by Andrade and shown to be of the type

$$\eta v^{1/3} = A e^{c/rt}$$

where v is the atomic volume, t the absolute temperature, and A and c are constants. Under some experimental conditions, however, more marked

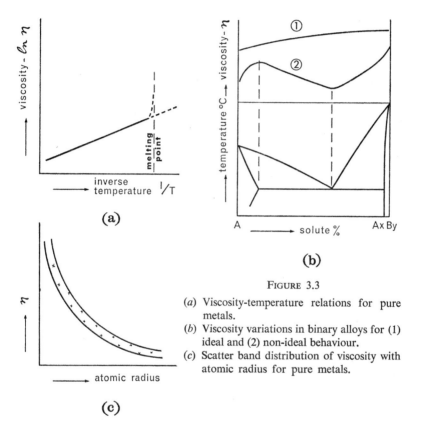

(a)

(b)

Figure 3.3

(a) Viscosity-temperature relations for pure metals.
(b) Viscosity variations in binary alloys for (1) ideal and (2) non-ideal behaviour.
(c) Scatter band distribution of viscosity with atomic radius for pure metals.

(c)

changes in viscosity near the melting point have been observed, the theoretical or experimental significance of which is not yet fully clear.

Viscosity variations with composition indicate that in some binary systems appreciable variations in viscosity occur at compositions corresponding to the occurrence of an intermetallic phase in the solid state. Viscosity data in this way corroborate the X-ray evidence of the existence of a non-homogeneous distribution of solute elements in the solvent metal of some alloys.

Viscosity studies have contributed to the understanding of the liquid state, but in general the scarcity of accurate viscosity data for alloys is responsible for the present lack of a comprehensive theory of the viscosity of metallic liquids.

3.2.3 Surface tension

Viscosity measurements indicate the nature of the weak interatomic forces holding atoms of a liquid together, but still allowing them a continuous random motion in a confined volume. The action of such bonding forces at the free surface of the liquid is to pull the surface atoms inwards owing to the non-symmetrical nature of the distribution of forces at the surface. The net effect of surface bonds is to cause a small volume of liquid to take up a spherical shape. The force P required to pull the surface of a liquid apart over the length l in the surface is defined as surface tension of the liquid σ. Per unit length

$$\sigma = \frac{P}{l}$$

Surface tension is measured by arranging an external and measurable force which will just balance the force of the surface tension in a given set of conditions. The surface tension of liquid metals can be found by measuring the rise of a liquid level in a capillary tube, or the contact angle of a drop of liquid lying on a flat surface, or the pressure necessary to detach a gas bubble from the tip of a capillary tube immersed in the liquid. While the experimental details and the equilibrium of forces equations for these methods are different, the principles can be briefly explained by reference to the gas bubble method as that most generally useful for liquid metals.

In order just to detach a gas bubble from the end of the tube (Fig. 3.4), it is necessary to expand its volume, and the work done is equal to $P \cdot dV$. This work is equal to the increase in energy in overcoming the surface tension $\sigma \cdot dS$, where S is the surface of the bubble. Assuming that the bubble is a perfect sphere of radius r and equating the above two energies, we obtain

$$P \cdot 4\pi r^2 \cdot dr = \sigma \cdot 8\pi r \cdot dr$$

or
$$P = \frac{2\sigma}{r}$$

The value of P under the experimental conditions used must also satisfy the equilibrium of forces; hence P_g, the total gas pressure on the manometer, is

$$P_g = \frac{2\sigma}{r} + \rho g h$$

The expression $\rho g h$ is the hydrostatic pressure in the liquid whose density is ρ, g is the gravitational constant and h is the distance of the bubble below the liquid surface.

The true value of σ is that which obtains when the liquid is in equilibrium with its own gas. Experimentally this can be measured in very few instances only, and for metals the measured values are frequently influenced by the presence of an adsorbed gaseous liquid or solid film at the surface. For this reason it is always important in using values of σ to note the nature of the gaseous atmosphere under which the value of σ was determined.

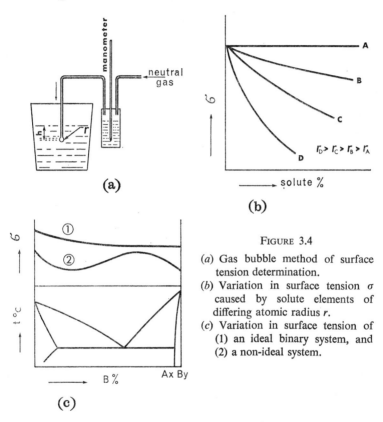

(a)

(b)

(c)

FIGURE 3.4

(a) Gas bubble method of surface tension determination.
(b) Variation in surface tension σ caused by solute elements of differing atomic radius r.
(c) Variation in surface tension of (1) an ideal binary system, and (2) a non-ideal system.

Studies of surface tension of liquid metals have been particularly extensive with respect to alloy constitution and to the effects of different solute elements on the surface tension of a given solvent. Surface tension and binary composition diagrams show that σ values fall smoothly over the whole composition range in a simple eutectic system, but deviations from a smooth curve occur in some systems with intermetallic phases. The effect of solute elements on the value of a given metal is shown in Fig. 3.4.

An important theoretical significance of the value of σ is that it helps in establishing a relationship between various other physical properties of liquids. In turn such relationships can sometimes be used to calculate the values of σ. The applications of σ in casting metallurgy are very wide in dealing with problems in refining of metals, fluidity, nucleation and growth phenomena, flow of metals and feeding of castings. The magnitude of some physical properties of liquids is indicated in Table 3.1.

3.3   PHYSICO-CHEMICAL PROPERTIES OF LIQUID METALS

3.3.1   Aspects of physico-chemical methods

Physical chemistry is concerned amongst other things with a range of properties of liquid metals, and some of these, like chemical reactivity which is important in metal melting and composition control, have been introduced in Chapter 2. A special aspect of this field, namely gas–metal reactions, is discussed separately in Chapter 4 because of its general importance.

Another field of physical chemistry which is important for melting and casting of metals is that which deals with the laws governing the co-existence of alloy phases at different temperatures and pressures. The most widely applied is the phase rule, which leads to the concept of the equilibrium diagram. In melting and casting of metals some aspects of equilibrium diagrams arise that are not normally discussed fully in the standard textbooks of physical chemistry and physical metallurgy. Consequently some general features of equilibrium diagrams will be briefly introduced, but their specific applications in founding will be considered at greater length.

3.3.2   General characteristics of equilibrium diagrams

For application in casting metallurgy an equilibrium diagram may be considered as a graphic summary of mainly experimental data showing the existence of physico-chemical units or phases in terms of alloy concentration and temperature. Such phases can be conveniently identified in the solid state by metallographic examination (polishing, etching and visual or microscopic observation). The phases are described as crystals or grains or constituents of the structure which vary in size (macro $>$ 0·1 mm or micro $<$ 0·1 mm), in shape (polyhedral or irregular outer surfaces) and in distribution (homogeneous or heterogeneous). In some alloys a number of phases can occur intimately mixed together; such mixture groups are also described as constituents (eutectics and eutectoids). The term 'complete cast structure' is discussed more fully in 6.7.1.

In a binary diagram, for example, the alloy concentration in weight or atomic percentages is plotted along one axis and the temperature along the other, indicating the fields of existence of different liquid and solid phases. One of the binary diagrams most widely used in practice is that of iron and carbon shown in Fig. 3.5, which can be used as an example for similar discussions of other binary diagrams.

The liquidus line marks the temperature—composition boundary above which iron and carbon exist as a completely random and homogeneous liquid solution, although possible departures from an ideal distribution of solute over the whole composition range have been indicated in preceding paragraphs. Above a certain carbon content this liquid becomes unstable and the carbon atoms separate out as graphite, so that this particular binary alloy exists for practical purposes up to a limited carbon content only.

When the temperature falls to the liquidus line at a given carbon content, for example X in Fig. 3.5, some solid forms, and solid and liquid phases co-exist until the vertical drawn through this carbon composition intersects the solidus line, indicating on the ordinates the actual temperature at which the solidification is completed under the equilibrium conditions. The solid formed for alloy X is of a structural type known as Fe_γ, or the austenite phase, which is only stable at higher temperatures. The breakdown of the austenite phase into Fe_α, or the ferrite phase, occurs at the temperature where the vertical for alloy X intersects the line separating the γ and α fields. The change into the α phase continues until 721°C is reached, at which temperature another phase forms from the remaining undecomposed γ phase. As the new phase Fe_3C (iron carbide) forms, it depletes the γ phase of carbon, giving the opportunity for more α phase to form. At a slow cooling rate both phases, Fe_3C and α, grow together, giving a fine mixture of the two phases, this constituent being described as pearlite.

This example shows the general simplicity of phase-temperature-composition interpretation of an equilibrium diagram. From any fully annotated equilibrium diagram the following main information can be readily obtained:

(a) For an alloy of a given composition the temperature is shown at which phase changes would occur under very slow (i.e., equilibrium) cooling or heating rates.

(b) At temperatures at which phase changes are occurring the diagram shows the composition of the new phase in equilibrium with the one that is beginning to transform.

(c) For all the fields where two phases co-exist the diagram can be used to obtain the relative amounts of the two phases under equilibrium by using the lever rule.

Although in the diagrams for three (ternary diagrams) or more metals a slightly more complicated plotting is used, the main principles of binary diagrams still hold. Furthermore a vertical section through a ternary diagram can be presented in the same way as a binary diagram (but cannot be interpreted quantitatively).

3.3.3 Application of equilibrium diagrams in casting metallurgy

Equilibrium diagrams are constructed on the basis of equilibrium or such slow rates of phase transformations that all the changes are allowed to proceed

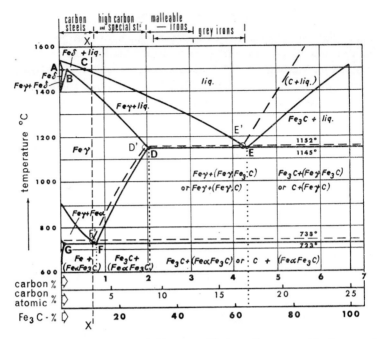

TERMINOLOGY:

Phases (crystals or grains)	Metallurgical term	Mixture of phases (constituents)	Metallurgical term
Fe_γ Fe_α Fe_3C C	austenite ferrite cementite graphite	(Fe_α, Fe_3C) (Fe_γ, Fe_3C) (Fe_γ, C)	pearlite (eutectoid) ledeburite (eutectic) graphite (eutectic)

MAIN COMPOSITION, C% BY WEIGHT:

A 0·08   B 0·16   C 0·53   D 2·03   D′ 2·01   E 4·30   E′ 4·26

F 0·80   F′ 0·68   G 0·025

INDUSTRIAL Fe–C ALLOYS:

Carbon steels contain regularly Si, Mn, P, S.

Malleable and cast irons are Fe–C–Si alloys containing regularly Mn, P, S.

Other unspecified elements also present in all industrial Fe–C alloys. Phases occurring in the eutectic range are strongly affected by cooling rate and impurities present.

to completion. For studies of cast metals, however, non-equilibrium cooling and heating rates are generally used as these are more closely related to the practice of making castings. It is therefore necessary to predict those modifications in the phase–temperature–composition relationships shown in the equilibrium diagrams which are likely to be produced by non-equilibrium transformation rates. Examples of such transformations will be considered in more detail in the discussion of both solidification and heat treatment phenomena, but reference to Fig. 3.5 may illustrate the importance of the non-equilibrium problem. The Fe–C equilibrium diagram anticipates that at room temperature the equilibrium phases should be ferrite and graphite. In fact the rate of cooling required to give these two phases at room temperature is so slow that an alloy containing ferrite and graphite phases can be obtained only under special conditions and when the carbon content is greater than 2%. Using normal cooling rates, ferrite and Fe_3C (carbide) phases are stable over the whole carbon range shown in Fig. 3.5. However, iron–carbon alloys containing more than about 2% can be readily made graphitic by additions of certain alloying elements, such as silicon, to obtain a family of widely used alloys known as grey or graphite irons. At lower carbon contents, e.g., in carbon steels, the theoretical instability of the Fe_3C phase is of little practical significance, apart from high temperature application of these alloys.

Another problem which is frequently met in the application of the equilibrium diagram to casting metallurgy is the interpretation of alloy composition. This can be illustrated with reference to the iron–carbon system. Although in this diagram the elements are indicated as Fe and C, the underlying assumption of metal 'purity' used in establishing the diagram may be significant. The concept of 'pure' iron or carbon has only theoretical significance, as already discussed in 2.4.2. Equilibrium diagrams are generally based on the highest purity metals available at the time the diagrams were established, but in casting metallurgy equally high purity metals are seldom if ever used. A normal steel or cast iron, for example, contains in addition to iron and carbon also silicon, manganese, phosphorus and sulphur, while many other elements may be present in smaller amounts.

The problems encountered when applying equilibrium diagrams for binary alloys to actual systems containing certain amounts of other elements as well are very varied. Such impurities may cause the appearance of a phase not predicted by the equilibrium diagram, as with iron–carbon alloys where small amounts of impurities may favour the presence of either Fe_3C or graphite. More frequently impurities influence the size, shape or even the amount of predicted phases (Chapter 6).

In the case of non-equilibrium transformations, equilibrium diagrams can be used to indicate the direction of the changes, and some of the consequences of the non-equilibrium rates. Examples of this kind will be discussed in

Chapters 6 and 7. In many instances the kinetic or rate factor superimposed on the equilibrium diagram may necessitate a construction of a special type of diagram for a clearer presentation of the kinetic effects. For example, very fast cooling of Fe–C alloys from the austenite range (Fig. 3.5) leads to a number of different transformation products, which are best explained in special diagrams, the so-called time–temperature transformation or S-curves for steels. This is an example of heat treatment further discussed in Chapter 6.

The equilibrium diagram is the starting point for analysing the presence and number of phases in the structure of cast alloys. With the additional evidence provided by microscopic, X-ray and other physical or chemical examination, departures from the predicted nature of the structure can usually be satisfactorily explained. The diagram does not however indicate the likely physical size or geometrical shape and distribution of various phases in the cast structure. These are determined by solidification and heat treatment phenomena, Chapters 5 and 6. Equilibrium diagrams are also indirectly useful for correlating casting and several other properties of alloys with their composition. Examples of such property diagrams are given in Figs. 3.3*b*, 3.6*c* and 3.7*b*. They are a valuable means of comparing a specific property behaviour with the composition of a given alloy system, as well as of comparing such properties for different alloy systems. Property diagrams are therefore of theoretical as well as practical interest.

3.4 CASTING PROPERTIES

3.4.1 General consideration of casting properties

It has already become apparent from the discussion of physical properties that alloys of certain compositions may have distinctly advantageous properties for use in casting processes. This implies that casting processes can be more readily carried out with such alloys, and also that castings of relatively higher properties, or quality, are likely to be produced. Some of these properties, e.g., mould filling ability (fluidity), low volumetric contraction during freezing, insensitivity to casting stresses or to segregation phenomena in the cast structure, are all of general interest as they concern most alloys. Some other properties are more specific and apply with greater emphasis to certain alloy compositions only. A comprehensive list of all the casting properties would be very long, and therefore only some typical general and specific properties will be introduced in the following paragraphs to illustrate the main principles involved. Other casting properties and tests are discussed in Chapters 4 and 6, and more detailed information regarding specific casting properties of various alloys is to be found in the lists of further reading.

The various casting properties and the respective tests which are encountered in foundry practice and described in the technical literature fulfil two main purposes: the development of casting alloys and casting processes, and various kinds of production control at the different stages of the casting process (Chapter 11). A few of the casting properties and tests are also used for more fundamental studies of alloy behaviour where physical or chemical measurements are either too complex or too difficult to carry out.

3.4.2   Fluidity

It has been generally observed in foundry practice that in filling moulds of intricate design, particularly those which include thin sections, everything else being constant, some alloys fill the mould cavity and reproduce its details in the finished casting better than others. This particular property of an alloy has been defined by foundrymen as casting fluidity or just simply fluidity. It is clear from this definition that fluidity, unlike the other properties already discussed in this chapter, is a complex empirical property. Experimental data indicate that fluidity can be influenced by certain physical and chemical properties of the alloy, by characteristics of the moulds, and by the methods used in filling the moulds. It is therefore necessary to design a test that will satisfy the empirical definition of fluidity, and then to relate such arbitrary fluidity measurements to the physical and physico-chemical properties of the metal and mould, and to the forces acting on the metal during mould filling.

Figure 3.6a shows the typical situation of a casting where a liquid metal is poured and fills the mould under the force of gravity starting with the potential energy of position H. At the entry into the mould the liquid metal has gained a certain amount of kinetic energy, $\dfrac{mv^2}{2}$, which is used in spreading the metal along the mould cavity and filling up the mould volume. At the same time the liquid metal is being cooled by the loss of heat to the mould walls, $\dfrac{\partial Q}{\partial t}$, so that in the course of mould filling the metal may reach solidification temperature at points far removed from the position of metal entry into the mould. It would theoretically be possible to treat such a fluidity problem in terms of a general energy equation taking into account the flow energy system and the heat energy losses during the flow. The constants of such an equation would be related to the thermal properties of liquid alloys and to physical properties, such as viscosity and surface tension, in order to account for the specific fluidity behaviour of different alloys. Such an approach to a quantitative fluidity evaluation has not so far proved fruitful in practice, mainly for lack of data on the physical and thermal properties required. Another difficulty in such a fundamental approach to fluidity testing is the lack of agreement amongst metallurgists in defining a fluidity standard test piece which could be used for experimental evaluation of the general fluidity energy equation. At the present time fluidity is measured by using a variety of different designs of test piece, with the object of obtaining a comparative measurement of the sensitivity of a particular type of fluidity test piece to different metal, mould and pouring variables.

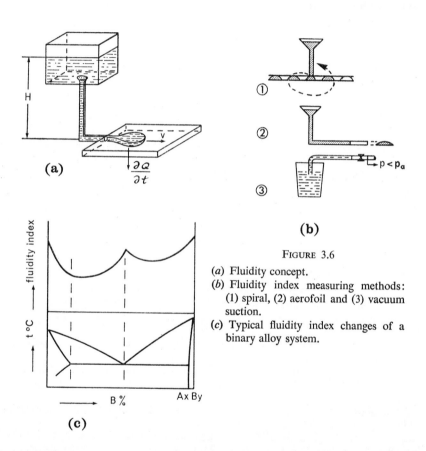

FIGURE 3.6

(*a*) Fluidity concept.

(*b*) Fluidity index measuring methods: (1) spiral, (2) aerofoil and (3) vacuum suction.

(*c*) Typical fluidity index changes of a binary alloy system.

3.4.3 Fluidity applications

A number of tests have been proposed for fluidity measurements, some of which are illustrated in Fig. 3.6. The problem in selecting a test piece is essentially that of using a specific feature in its design to show a measurable degree of sensitivity to the variables of special interest for a particular application. The standard unit or fluidity index which is usually chosen in various tests is the length or the area of the test piece obtained under constant controlled conditions of making the tests. Typical test results obtained by several kinds of fluidity test are shown in Fig. 3.6c. Such results reveal, for example, that an arbitrary fluidity index of an alloy increases with the pouring temperature of the alloy, and that, under otherwise comparable test conditions, fluidity values of a pure metal and of a eutectic alloy are greater than those of alloys solidifying over a temperature range. This observation helps to explain one of the reasons for the wide practical preference for eutectic or near eutectic alloys for founding purposes, particularly for castings with thin sections. Such tests may also be used for comparative studies of the effect of mould variables on fluidity, or for the control of casting quality in foundries (Chapter 11). On the basis of fluidity data for binary or more complex alloys, it is possible to make a quantitative selection of suitable alloys for specific moulds, casting designs or casting processes. For example, the useful effect of adding phosphorus to grey irons for production of thin section castings can be deduced and explained from fluidity—composition graphs. In this instance phosphorus increases the proportion of eutectic in the alloy as well as lowering the freezing temperature; both factors improve the degree of mould filling. Similarly the effects of mould dressings on mould filling can be explained in terms of fluidity test data (Chapter 8).

In fluidity applications of this kind it is essential to use the most appropriate test method. For example, the standard fluidity spiral test is too insensitive to show the effect of sodium in lowering the fluidity of aluminium–silicon eutectic alloys, but this effect is confirmed in foundry practice and it could be revealed by using fluidity tests sensitive to surface tension variations (e.g., fluidity index measurement based on the filled area of the test piece).

Fluidity data, in the absence of a general fluidity equation, are at present obtained and interpreted separately for each particular practical problem; e.g., the effect of pouring temperature can be analysed in terms of available heat and heat transfer in the mould. The effect of alloy composition on fluidity is more difficult to explain quantitatively, since—apart from the possible heat content variation with composition in a binary alloy—other physical properties such as viscosity and surface tension alter at the same time. The available experimental evidence suggests that the major factor affecting fluidity variation with composition in an alloy is the mode and rate of growth of the crystals during solidification; in other words, the manner in which the solidifying skin of the casting forming away from the mould wall presents a growth interface which may hinder the flow of the liquid in filling the mould. The hindrance caused by crystals with irregular growth surfaces (dendrites) in long freezing range alloys is much greater than that of the comparatively smooth crystallisation interface front of pure metals and eutectic alloys. This feature of crystal growth morphology is further discussed in Chapters 5 and 6.

3.4.4 Volume expansion and contraction (shrinkage)

Concepts. One of the general properties of solids or liquids in relation to their heat energy content is the increase or decrease in their dimensions due to the absorption or loss of heat. This is caused by an increase or decrease in the amplitude of atomic vibration in the crystal structure. Such dimensional changes are usually expressed in terms of linear or volumetric expansion or contraction coefficients, which have characteristic numerical values for different substances. At the melting point a large increase in the heat absorbed (latent heat of fusion) leads to the breakdown of the solid state bonding, and

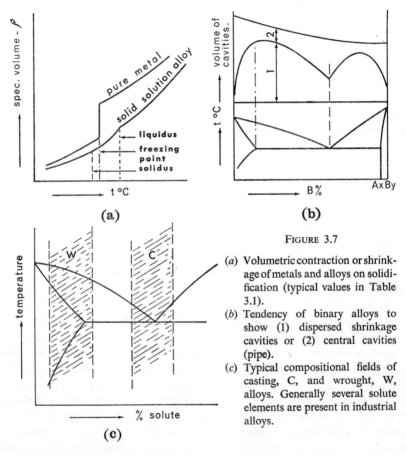

FIGURE 3.7

(a) Volumetric contraction or shrinkage of metals and alloys on solidification (typical values in Table 3.1).

(b) Tendency of binary alloys to show (1) dispersed shrinkage cavities or (2) central cavities (pipe).

(c) Typical compositional fields of casting, C, and wrought, W, alloys. Generally several solute elements are present in industrial alloys.

a discontinuous increase in volume is generally observed. Such volume expansion on melting is general, but a few metals such as bismuth and antimony anomalously contract. A further increase in volume expansion occurs in the liquid state, and the whole process is reversed on cooling. The problem of linear expansion on heating is of importance in the heat treatment of castings, but the contraction phenomena on cooling from the liquid state are of wider interest and are generally described as the shrinkage or contraction phenomena in cast metals.

Shrinkage properties. Volumetric changes on cooling a pure metal or an alloy from the liquid state can be summarised graphically as shown in Fig. 3.7a. Because of the experimental difficulties (similar to those of measuring other properties of metallic liquids), accurate data on numerical values of volumetric liquid and solidification contraction are available for some metals and alloys only. For general guidance the volume contraction characteristics of some pure metals and alloys are summarised in Table 3.1.

TABLE 3.1 *Approximate values of some physical and casting properties of liquid metals and alloys near their freezing temperatures*

Metal and alloy system	Composition, wt %	Viscosity, centipoises	Fluidity, arbitrary units 100 for pure metal	Surface tension, dynes cm^{-1}	Density, g cm^{-3} at 20°C and liquid temperatures	Volume change on solidification, % of value at 20°C
Al	99.99	2·84	100	860	2·69 2·38	6·6
Al–Cu	Max. solid solubility (5·7)	3·1	60	850	2·80 2·48	6·3
Al–Cu	Eutectic (33)	2·4	125	—	3·45 3·14	4·8
Al–Si	Max. solid solubility (1·65)	3·4	65	820	2·68 2·43	6·2
Al–Si	Eutectic (11·7)	1·8	68	820	2·65 2·46	3·5
Fe	99.99	6·3	100	1790	7·87 6·79	3·5
Fe–C	Max. solid solubility (2·01)	8·5	70	1550	7·75 7·05	—
Fe–C	Eutectic (FeC 4·26) (Fe$_3$C 4·3)	3·4	140	1330	7·22 7·16	Fe–C, 0–1 Fe–Fe$_3$C, 3·4

The main feature emerging from an examination of contraction data of metals and alloys is that the values are dependent on composition and phase constitution, and differ considerably from one metal to another. In a few alloys, for example cast iron, the magnitude of solidification contraction is not constant, but depends on the nature of phases which form during the solidification. Due to its large specific volume, when graphite forms on freezing, the volumetric solidification contraction is small or zero, but with the Fe_3C phase crystallising instead this contraction is larger.

Shrinkage measurement. Evaluation of expansion or contraction coefficients in the solid state presents few experimental difficulties. Solidification and liquid state volume measurements and changes require specialised techniques. The method generally applied is that based on the Archimedes', or immersion, principle, using the liquid metal contained in a crucible immersed in a special liquid. The immersion liquid is usually made up of suitable inorganic mixtures (salts or oxides), and the crucible material must be inert to the liquid metal as well as to the immersion liquid. If such liquids are not available, volume changes are measured directly in containers of special design by the picnometer method.

Shrinkage applications. Unlike fluidity, specific volume and volumetric shrinkage are well-defined physical properties, and one would expect a concise and quantitative application of such quantities in studies of cast metals. That this is not so in practice is mainly due to the nature of the industrial problems created by volumetric shrinkage phenomena. These problems fall into three main groups: (i) those arising in the production of castings of high dimensional accuracy, (ii) those concerned with controlling the solidification process in such a way that no cavities result in the casting from volumetric contraction in the liquid and during solidification, and (iii) those which arise during contraction or expansion in the solid state. The nature of these problems is outlined below. In general, for a practical solution of some shrinkage problems, knowledge of the absolute values of volumetric contractions is not essential (Chapter 9).

When a mould containing a cavity of a given volume is filled, the liquid metal begins to contract on cooling. This decrease in volume can be readily compensated by adding more liquid metal to the mould to keep it full, until the solidification temperature is reached. Thus, in general, liquid metal contraction should present few problems in casting practice. In some cases special precautions have to be taken to ensure that all parts of the mould are full; for example, the air must be able to escape from the mould cavity or an incomplete or undersized casting results. During solidification, however, the problem of supplying liquid metal to all parts of the mould where volume contraction is taking place may be complex. Many parts of the casting could solidify with cavities as a result of incomplete liquid metal compensation for the volumetric contraction. Thus, in the solidification problem the question is not that of the magnitude of the freezing contraction, but rather that of ensuring that solidification occurs in such a way that none or only a small proportion of volume contraction is left in the casting as cavities or porosity of various shapes, sizes and distribution. Shrinkage cavity sensitivity of a binary alloy can be shown in a property diagram, Fig. 3.7b, where the volume of internal shrinkage cavities is plotted against the alloy composition. Shrinkage sensitivity as a casting property is defined as the volume of internal cavities likely to result if a casting of an arbitrary design is produced in an identical manner from all the alloys examined. The origin and significance of this diagram will be discussed more fully in Chapters 6 and 10.

The contraction problems in a solid casting as it cools from the solidus to room temperature are of a different kind. They originate usually from non-uniform temperature distribution, and hence non-uniform contraction in the casting during or after solidification, or they are caused by the fact that such a contraction may be hindered by the mould and casting design. In general, the linear contraction of the solid casting is non-uniform, Chapter 6.

As a result of non-uniform contraction in the solid state two major problems of the casting process arise. Firstly, non-uniform contraction gives rise to various types of stress in the casting (6.10) which can lead to one or more of the following: non-uniform plastic deformation and hence possible warping of the casting, residual stresses in the casting, or fracture (cracking). Stress sensitivity of an alloy is a measure of its ability to be cast into shapes where the stress problems are particularly severe. Results of various types of test used can be plotted in a property diagram, an example of which is shown in Fig. 6.10c.

Another consequence of solid contraction which is of major practical significance is that the room temperature dimensions of a casting differ to a smaller or larger extent from those of the mould cavity or the pattern used. In order to obtain a casting of required dimensions, it is therefore necessary to enlarge the size of the patterns. The required enlargement is defined in mineral mould practice as the patternmaker's shrinkage allowance. Due to the non-uniform nature of solid contraction of a casting, the problem of evaluating shrinkage allowances accurately is difficult, particularly for castings of complex design. Empirical rules which are used for such purposes are discussed in reference books on cast metals.

3.4.5 Specific casting properties and tests. Melt quality

The general casting properties discussed so far, although complex and involving metal, mould and process variables in their evaluation, apply nevertheless to most alloys. Some other casting properties and their tests are used in founding with a more limited objective for dealing with some specific problem of a given alloy or of a casting process. These properties and tests are used as a rule for process control (Chapter 11), and the most typical examples, known as melt quality tests, are summarised in the following

3

paragraphs. Others, notably the mechanical property tests, are dealt with in Chapter 11.

Composition tests. These tests are typical amongst metal quality tests. Liquid metal in a furnace is of complex composition containing the required elements and, in addition, unspecified elements (impurities) some of which may be more harmful than others to the castings produced. Consequently many tests are used in foundries to reveal liquid metal 'quality' as a substitute for a full chemical analysis, which may be either too costly or not available quickly enough. Melt quality tests are therefore used to demonstrate the presence or absence of those elements in the liquid metal which may be critical for casting quality. Some such tests relate to gases in metals, and will be discussed in Chapter 4, while others concern the specified alloy composition. The most important amongst such tests are those relating the cast structure of an alloy to its composition. If the cast sample is cooled in a reproducible and standard manner, then for many alloys a useful correlation can be obtained between alloy composition and its micro- or macro-structure. Such a test can be used for approximate composition control (with, for example, manganese brasses, cast irons and carbon steels), but qualitative tests of this type are gradually being replaced with quantitative analytical control tests (Chapter 11). With cast irons, taking a cooling curve of a melt sample can be used as a quick and approximate method of obtaining an indication of the content of alloying elements promoting graphite formation in the cast structure (C.E. %), 7.6.3. This is an example of a more unusual application of the equilibrium diagram.

Shrinkage tests. These are of two types: those used for comparing different alloys with respect to their shrinkage sensitivity, as discussed in 3.4.4, and those used to compare shrinkage sensitivity variations as a result of process variables for one particular alloy (metal quality shrinkage tests). This latter type of test indicates that some compositional variations in melting accompanied by a change in the solidification behaviour may alter the shrinkage sensitivity of an alloy. The principle used is to induce a cast test piece to solidify unidirectionally, so that most of the shrinkage behaviour becomes apparent at the free surface of the test piece. Visual or calibrated observations against set standards enable deductions to be made about the departure from the required composition and how this may be corrected. The best practical illustration of such a test is the tough pitch copper test. The flatness of the top surface of the cast sample can be directly related to the hydrogen–oxygen content in copper. A flat top surface in the solidified sample implies a correct proportion of the gaseous elements in copper; an excess of oxygen leads to surface sinking, while an excess of hydrogen is accompanied with top surface cauliflowering (Chapter 4).

Cast structure and other property tests. Various special tests can be designed to test the structure of certain alloys. These are used as melt quality tests prior to the pouring of castings, in order to avoid possible production of castings with an unsatisfactory cast structure. For example, it may be important to ensure that cast iron microstructure is of a particular type. The microstructure of most cast irons is sensitive to composition, melt treatment and the cooling rate of the casting. Consequently, special tests, some of which are shown in Fig. 3.8, relate composition and cooling rate

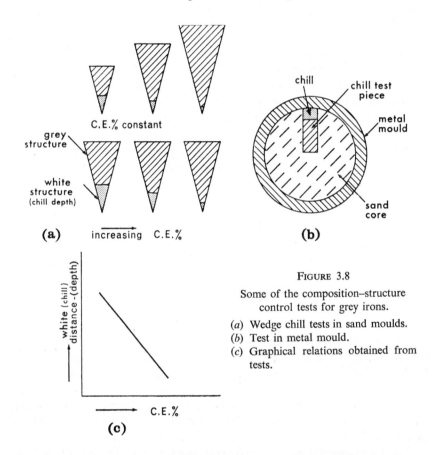

FIGURE 3.8

Some of the composition–structure control tests for grey irons.

(a) Wedge chill tests in sand moulds.
(b) Test in metal mould.
(c) Graphical relations obtained from tests.

effects to the microstructure of cast irons. In the wedge test, for example, the depth of the thin zone which solidifies white can be related to the composition of grey iron by using the concept of percentage carbon equivalent, C.E.% (Chapter 7), and predicts the suitability of the particular grey iron for producing castings of a given cross-section and properties. When using grey iron for piston rings, the structure and the residual stress in a ring are important; the latter is tested by measuring the gap width formed when cutting through a ring (stress tests).

3.5 CASTING ALLOYS, WROUGHT ALLOYS AND ALLOY SELECTION

Alloys suitable for founding processes are referred to as casting alloys. In addition to the casting properties already discussed and which particularly concern the metallurgical behaviour in the liquid state, during and after the solidification, a more complete definition of a good casting alloy would also include the engineering and production considerations at all stages of the casting process, starting with the raw materials and ending with the performance of the casting in its application. The major factors would be availability of raw materials, ease of melting and composition control, freedom in choice of casting design and suitability to deal with dimensional accuracy, soundness, surface and stress problems in casting, ease of controlling and varying the cast structure, satisfactory behaviour in the subsequent manufacturing processes (machining, joining and welding), good combination of useful properties for diverse applications (stress, temperature and corrosion) and last but not least the overall cost.

In theory, any cast shape could be made in an alloy of any composition and by any one of a large number of casting processes. Technical and economic factors restrict such a free choice, so that in practice alloys having the best combination of casting properties are used for specific applications and castings are made by the most appropriate process. In general, sand and mineral mould processes impose the least number of restrictions on alloy behaviour in casting, and there is therefore a wide choice of casting alloys in this field to produce the various properties of castings. Metal mould processes for shaped castings, on the other hand, impose severe restrictions, particularly on mould filling and stress behaviour, and consequently the choice of suitable alloys for these processes becomes more narrow. The restrictions increase further when changing from gravity to pressure die casting. The solution of the problem of selection of an alloy for a casting process requires therefore a joint consideration of metallurgical, engineering and production requirements. This problem is further considered in Chapter 7 (7.7).

The term wrought alloy implies that some alloys are specially advantageous for the wrought or fabrication processes of manufacture. As wrought processes start with ingots, which are primarily castings, it follows that the casting properties previously discussed must apply to an ingot casting process. This is in fact so, and, though ingots are cast into metal moulds, the basic geometrical simplicity of ingot shape imposes far fewer restrictions on the choice of composition of ingot casting alloys in comparison with the alloys used for shaped castings. The more significant properties defining wrought alloys are, firstly, those that control their plastic behaviour in subsequent working of the ingots and, secondly, those that determine their suitability for various applications. Both of these groups of properties are dependent on the structure of an alloy (structure-sensitive properties, Chapter 7). Experience has shown that the best combination of properties for a wrought alloy (for a simple binary alloy) is encountered in the field of composition, W in Fig. 3.7c. Similarly, the best combination of casting and other properties for foundry alloys is usually around the eutectic range of composition, C in Fig. 3.7c. However, in some cases, the contribution of production, engineering or some other specific property factors may swing the pendulum in one or other direction, and some casting alloys used in the industry are compositionally in the field 'W' and wrought alloys in field 'C'. This further emphasises the point that often a large number of factors have to be considered in the selection of an alloy for a given process and for a given application. In other cases, the uniqueness of an alloy or a process or a casting design may narrow the range of choice to one or two alloys only.

3.6 SUMMARY

The physical, in comparison with the chemical, behaviour of liquids is not theoretically so well understood. Some of the physical properties of liquids of major interest in metallurgy are difficult to measure. In the absence of such numerical data and relevant theories, many of the problems arising in the casting of metals, such as fluidity or contraction phenomena, are dealt with in practice by empirical and direct measurement tests. A major exception to empirical rules is the range of these problems arising in the solidification of alloys, which can be related to equilibrium diagrams. Although the alloys used industrially are very complex and their structure departs from the equilibrium conditions of solidification, equilibrium diagrams are the starting point of the discussion of cast structure.

FURTHER READING

FROST, B. R. T., The structure of liquid metals, *Progr. Metal Phys.*, 1954, **5**, 96.
WILSON, J. R., The structure of liquid metals and alloys, *Metall. Rev.*, 1965, **10**, No. 40.
ANDRADE, E. N. DA C., A theory of viscosity of liquids, *Phil. Mag.*, 1934, **17**, 497.
YAO, T. P., Über die Viskosität der metallischen Schmelzen, *Giess., Tech-wiss. Beihefte*, 1956(16), 837.
SEMENCHENKO, S. T., *Surface Phenomena in Metals and Alloys*, Pergamon Press, Oxford, 1961.
BASTIEN, P., J. C. ARMBRUSTER and P. AZOU, Fluidity and viscosity, *Trans. Amer. Found. Soc.*, 1962, **70**, 400.
RAGONE, D. V., C. M. ADAMS and H. F. TAYLOR, Fluidity, *Trans. Amer. Found. Soc.*, 1956, **64**, 653.

COOKSEY, C. J., V. KONDIC and J. WILCOCK, The casting fluidity, *Br. Foundryman*, 1959, **52**(9), 381.

DESCH, C. H., Physical factors in casting of metals, *Proc. Inst. Br. Foundrymen*, 1936–37, **30**, 77.

GOODRICH, W. E., Volume changes during the solidification, *Trans. Farad. Soc.*, 1929, **25**, 531.

CHAPTER 4

GASES, LIQUID METALS AND CASTINGS

4.1 INTRODUCTION

Gas–metal relationships are of greater importance in the melting and casting of metals than in any other branch of applied metallurgy. The main reason for this is that liquid metals are in contact with a gaseous atmosphere during melting and to a smaller extent during casting; at the high temperatures involved, one or more of the furnace atmosphere or mould gases may react with the liquid metal. Gases can perform a temporarily useful function in process metallurgy, as in oxygen blowing in the refining of steels, but in the majority of foundry processes gas contamination of liquid metals is undesirable. The basic laws governing gas–metal chemistry are the same as those that apply for liquid metal solution systems, but there are certain detailed features that deserve separate consideration. In this chapter the emphasis will be on the metallurgical problems arising in the application of gas–metal laws, but some of these will first be briefly summarised.

4.2 PHYSICAL CHEMISTRY OF GAS–METAL SYSTEMS

4.2.1 Principles of solubility and compound formation

In the same way that one metal may dissolve in another, certain gases may dissolve in certain metals. Studies of the driving forces leading to the solution of gases in metals, the general laws of solution, the effects of temperature, pressure, and gas or alloy concentration on the solubility, are of special interest because of their wide metallurgical applications.

The fact that a metal dissolves a gas can be indicated by an equation, for example:

$$Al_{(l)} + H_{2(g)} \rightleftarrows 2\underline{H} \, *$$

showing that a molecule H_2 of hydrogen gas (g) dissociates in contact with the aluminium liquid (l) and goes into solution as atomic hydrogen $\underline{H}$. (The hydrogen gas originates mainly from dissociation of H_2O vapour in the furnace atmosphere.) On the other hand, in the system $Al_{(l)}$ and $O_{2(g)}$ the product of the reaction is a solid (s) compound:

$$Al_{(l)} + O_{2(g)} \rightarrow Al_2O_{3(s)}$$

---

* Reaction of this type can be conveniently discussed without complete stoichiometric notation.

57

Al_2O_3 is a solid film (under normal conditions of melting) which forms at the surface of the liquid $Al_{(l)}$. In other words, oxygen does not dissolve in a measurable order of magnitude in $Al_{(l)}$ but forms a solid film of oxide at the gas–metal interface. Suspended Al_2O_3 inclusions may be obtained in $Al_{(l)}$ under certain melting and casting conditions when the oxide film is mixed into the liquid, and the aluminium alloy melt may have to be cleaned by fluxes, usually made up of alkali chlorides and fluorides.

There is no simple chemical postulate which would summarise the conditions or predict quantitatively the extent of solubility or of compound formation between a given metal and a given gas. Present knowledge of the behaviour, as well as the quantitative treatment, of gas–metal systems is largely based on experimental data. However, where the basic data are available, the laws of physical chemistry can be applied to deal with some typical metallurgical problems.

The two different types of gas–metal reaction discussed above illustrate another general characteristic. The rate of the reaction between aluminium and hydrogen varies according to the temperature, pressure and concentration, but under normal conditions of melting this reaction is relatively slow, and the amount of $\underline{H}$ dissolved in $Al_{(l)}$ seldom reaches the equilibrium value, as determined by the laws of physical chemistry. The reaction between aluminium and oxygen, on the other hand, is so fast that equilibrium is reached almost instantaneously. This would suggest that a melt of liquid aluminium would oxidise rapidly and become converted fully into Al_2O_3. However, the film of Al_2O_3 which forms instantaneously at the free surface of $Al_{(l)}$ is continuous and prevents the diffusion of oxygen from the atmosphere which is necessary for the reaction to continue.

In contrast, with a magnesium melt the oxide forms as a powder, and $Mg_{(l)}$ continues to oxidise so rapidly that the liquid bath starts burning violently unless suitably protected by liquid fluxes or special atmospheres. These examples demonstrate that for the application of physical chemistry laws to gas–metal systems it is necessary to establish experimentally for each system the nature of the reaction (equilibrium laws) as well as the limiting factors controlling the rate at which it takes place (reaction kinetics).

4.2.2 Solution laws

Simple molecular gases: limited solubility. When a simple molecular gas, for example H_2 or N_2, dissolves in a given liquid metal $M_{(l)}$, then the solution reaction can be typified by the relation

$$M_{(l)} + H_{2(g)} \rightleftarrows 2\underline{H}$$

This indicates a process of dissociation of the molecular gas such as $H_{2(g)}$, and its dissolution as atomic hydrogen $\underline{H}$ in the liquid metal $M_{(l)}$. The reaction between a solid metal $M_{(s)}$ and a molecular gas such as H_2 proceeds in exactly the same way, but its rate in general is very much slower, so that in practice most gases occurring in solid metals have been absorbed while the metals were liquid. Some solid metal–gas reactions may proceed relatively quickly, particularly at high temperatures or when H_2 is present in the atomic state; an example of the application of this principle is acid pickling of metals. At a given temperature of the liquid metal $M_{(l)}$ and pressure p of hydrogen $H_{2(g)}$ above the melt, the amount of hydrogen dissolved $\underline{H}$ will reach a specific value when equilibrium is fully established and no further hydrogen absorption will occur. This equilibrium value of solution will change with the temperature of the metal and the pressure of hydrogen, p_{H_2}, above the melt, and such changes obey laws of equilibria summarised in a subsequent paragraph.

Simple molecular gases: extensive solubility. With some gas–metal systems pressure and temperature conditions are normally such that the type of reaction discussed in the previous paragraph operates. With other gases, particularly O_2 (or N_2), the reaction is similar for low values of oxygen concentration and pressure, but during solidification the dissolved atomic oxygen may separate as metal oxide. Hydrogen atoms, on the other hand, generally remain in solution in the solid metal, and metal hydride formation is relatively rare in alloys used in the industry.

As conditions leading to the solution of oxygen and to metal oxide formation are frequently encountered in metallurgical practice, metal–oxygen systems are studied not only by considering reaction relationships but also by making use of metal–oxygen binary equilibrium diagrams.

Compound gases. Furnace and mould atmospheres usually contain gaseous compounds, such as CO_2, CO, H_2O, SO_2 and hydrocarbons. Reactions between compound gases and liquid metals proceed in a similar way to those for the simple molecular gases; the compound gas dissociates at the liquid metal surface, and the atomic components of the gas dissolve in the metal or form a compound with it. For example, with H_2O we can have

$$M_{(l)} + H_2O_{(g)} \rightleftarrows MO_{(s)} + 2\underline{H}$$

or alternatively

$$M_{(l)} + H_2O_{(g)} \rightleftarrows \underline{O} + 2\underline{H}$$

The former holds for Al and H_2O systems, and the latter for Cu or Fe and H_2O systems. Other compound gas reactions are similar. As with the molecular gases, each metal-compound gas system has to be studied experimentally to obtain basic numerical equilibrium data, and to establish the factors controlling the reaction kinetics.

4.2.3 Gas–metal equilibria

The effects of pressure. The problem of the solubility of a gas in a metal can be considered in terms of a simple chemical reaction. At a given temperature and pressure when the reaction has proceeded to a state of equilibrium, that is to say no further unidirectional changes are occurring in the system, there is a constant ratio, defined as the equilibrium constant K, between the active masses of products and reactants.

For a simple molecular gas at a given temperature t,

$$M_{(l)} + H_{2(g)} \rightleftharpoons 2\underline{H}$$

and

$$K_t' = \frac{[\underline{H}]^2}{p_{H_2}}$$

which can be rewritten

$$\underline{H} = \sqrt{K_t' \cdot p_{H_2}} \quad \text{or} \quad K_t\sqrt{p_{H_2}}$$

This relationship, known as the square root or Sieverts' law for the solubility of elemental molecular gases in liquid metals, states that the gas dissolves in the atomic state and the amount of gas dissolved $\underline{H}$ is proportional to a constant K_t (which varies with temperature and alloy composition and the numerical magnitude of which depends on the units used) and to the square root of the pressure of H_2 gas above the liquid metal. If the value of K_t and the pressure of the gas p_{H_2} above the liquid metal—or its partial pressure when the hydrogen is a component of a gas mixture—is known, then the amount of gas dissolved in the metal in the equilibrium state can be calculated.

Some special examples of the application of this law in melting practice are discussed on pages 67 and 70 and in §4.4.1, page 74.

For a compound gas–liquid metal reaction the concept of equilibrium constant relationship is identical with the above example, but frequently different units are used for the active masses. For example, the reaction between liquid metal and water vapour can be expressed as:

$$M_{(l)} + H_2O_{(g)} \rightleftharpoons \underline{O} + 2\underline{H}$$

$$K_t = \frac{(\underline{O})\%(\underline{H})^2}{p_{H_2O}}$$

In this case the amount of hydrogen $\underline{H}$ dissolved in a molten metal can be calculated if the value of K_t, the pressure of the water vapour above the liquid metal, p_{H_2O}, and the oxygen concentration, $\underline{O}\%$, in the melt are known. Both the above examples emphasise the need for the correct use of the units for the *active masses* in the ratio expressing the value of the equilibrium constants. For the gaseous components above the melt, convenient units are pressures (any consistent system of pressure units); the components in solution may be expressed in terms of concentration per cent or per million by weight, atomic or molecular percentages or volume concentration, such as $cm^3/100$ g.

The effect of temperature. The value of K_t for a given gas–metal reaction varies with the temperature. In an endothermic reaction, where the solution of a gas in a metal is accompanied by an absorption of heat, the solubility increases with the temperature, while the opposite holds for exothermic reactions (Le Chatelier principle). Both are encountered with gas–metal systems; the solubility of $\underline{H}$ increases with temperature in liquid aluminium, magnesium, copper, iron, and many others, and decreases with temperature in titanium, zirconium, palladium, cerium and similar metals.

The practical usefulness of the equilibrium laws lies in the fact that many relationships can be derived between various properties of a system, which can be used to calculate or predict changes in any one quantity when the others vary. This can be illustrated by examining the variation of the

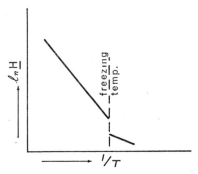

FIGURE 4.1 Endothermic solubility of a molecular gas, such as hydrogen, in a metal (see also Fig. 4.2b).

equilibrium constant with temperature. The equilibrium constant is related to the absolute temperature T by the equation

$$\frac{d \ln K}{dT} = \frac{\Delta H}{RT^2}$$

where ΔH is the heat of solution of a gas in the metal (this is assumed to be constant over a narrow range of temperature, a fact experimentally confirmed for several gas–metal systems), and R is the gas constant.

Using the square root law for K and assuming the value $p = 1$, by substituting in the above equation we find

$$d \ln (\underline{H}) = \frac{\Delta H}{RT^2} \, dT$$

which gives by integration

$$\ln \underline{H} = -\frac{\Delta H}{2RT} + C$$

This can be plotted to relate the solubility of a simple molecular gas, $\ln \underline{H}$, to the temperature $(1/T)$, as shown in Fig. 4.1.

If the heat of solution, ΔH, of a gas is known, the above relationships can be used to obtain either the solubility or the equilibrium constant of a gas as a function of temperature, or, alternatively, if the solubility of a gas is experimentally measured at various temperatures, then ΔH can be calculated for a given system.

Free energy calculations for gas–metal systems. It was pointed out in Chapter 2 that the course of a chemical reaction could be suitably followed by considering the free energy changes which occur. The selection and application of particular basic laws for dealing with applied problems frequently depend on the kind of measurements and properties that are used for verification of the laws or for studying the general behaviour of a system under the given conditions. With gas–metal systems direct measurement of solubility is generally the simplest approach for elemental molecular gases, while for compound gases either solubility or energy measurements of reactions are employed, although in principle either method can be used for both cases. For example, the reaction

$$Cu_{(l)} + H_2O_{(g)} \rightleftarrows \underline{O} + 2\underline{H}$$

can be analysed from free energy data, from which the equilibrium constant may be determined by using the relationship

$$\Delta G_0 = -RT \ln K$$

Alternatively, the solubility of $\underline{H}$ and $\underline{O}$ in liquid copper can be determined experimentally for a given pressure of H_2O and the equilibrium constant obtained directly from such measurements. Free energy analysis can be looked upon as a general approach, whilst the solubility method is often a short cut to a specific solution.

The effect of alloy composition. The presence of metallic or non-metallic solute elements in the alloy may profoundly affect the quantitative relationships of a binary gas–metal system. The behaviour of gas–metal systems does not normally follow the ideal laws, and the presence of a third element usually requires experimental confirmation before the data for binary systems can be extrapolated for ternary or more complex industrial alloys.

Physical problems. So far the discussion has been limited to the physical chemistry of gas–metal systems. A number of problems also arise that are of a distinctly physical nature. These include: the state of gaseous atoms in liquid metals; whether such atoms are ionised and, if so, what happens to the electrons of the dissolved gas; the nature of the dissolved gas; its distribution in the liquid or solid metal in the form of holes or cavities; inhomogeneity; interstitial or substitution types of solubility; the problems of diffusion; gas bubble formation from solution; lattice distortion of the solid metal; and effects of dissolved gas atoms on plasticity and residual stresses. The theoretical background to these problems may be studied in the texts referred to at the end of the chapter, but some of the more practical aspects are discussed in the following section.

4.3 PRACTICAL ASPECTS OF GAS–METAL SYSTEMS

The fact that metals are melted, poured into moulds, heat-treated and used, all in gaseous environments, explains the wide variety of gas problems encountered in the metallurgy of cast metals. These fall into four groups:

(a) The measurement and estimation of gas content.
(b) The methods of removal and control of gases in metals.
(c) The state and nature of gas occurrence in the structure of cast metals.
(d) The effects of gases on the properties of cast metals.

4.3.1. The measurement and estimation of gases in metals

Measurement of gases in metals may have two distinct purposes: (a) for the study of gas–metal equilibria, or (b) to determine the gas content of liquid metals in a melting furnace or in a solidified casting.

Equilibria measurements. Several types of apparatus are used for measurements of equilibrium gas content in metals, but most of them are built round the main concept, which is illustrated in Fig. 4.2a. This normally consists of a crucible with the liquid metal at a given temperature, pressure gauges, a vacuum pump, and a supply of the purified gas whose solubility is being investigated. The measurement consists of admitting the gas into the evacuated apparatus of a pre-calibrated volume, to build up a certain pressure, and then measuring the change in pressure until equilibrium is established. The drop in pressure indicates that the gas is being dissolved in the metal and its volume can readily be obtained by using the ideal gas equation. The apparatus for compound gas equilibria has additional features, and the evaluation of results may also involve the determination of the concentration of products of the reaction in the liquid metal. These are obtained by solidifying the sample after equilibrium has been reached, and analysing it by chemical or physical methods (Chapter 11).

The results can be presented in various ways: numerically or graphically as free energy data; converted into equilibrium constant equations or graphs; or presented graphically as solubilities for different temperature and pressure conditions. The last two methods have the most direct application in the metallurgy of cast metals and are summarised in Fig. 4.2. Figure 4.2b shows a typical example of a simple molecular gas solution in a metal. The solubility of the gas increases continuously with increasing temperature in the solid state, increases sharply at the melting point, and in the liquid state increases for endothermic and decreases for exothermic solutions. The units used for the ordinates in diagrams of this kind are frequently $cm^3/100$ g at normal temperature and pressure. The choice is a matter of convenience; the above units are suitable for most practical purposes, but they can readily be converted into more suitable units for free energy calculation.

In compound gas reactions one gas only may be dissolved, which can be dealt with in the way outlined above; an example of this is when hydrogen

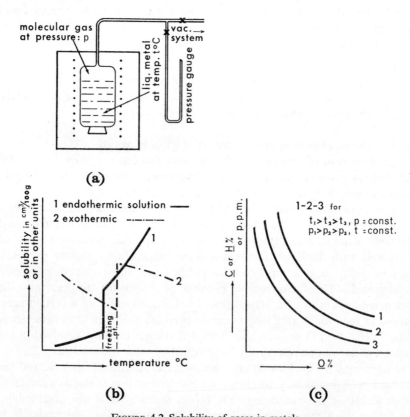

(a)

(b) **(c)**

FIGURE 4.2 Solubility of gases in metals.

(a) Method for equilibrium measurement of solubility of molecular gases.
(b) Endothermic and exothermic solubility of molecular gases.
(c) Compound gas relationship at constant pressure or temperature.

from water vapour dissolves in aluminium. When both products of the reaction go into solution, then for practical purposes the relationship between the dissolved products is useful. A high concentration of one product of the reaction leads to a low concentration of the other. Such a relationship holds, for example, for the solution of $\underline{H}$ and $\underline{O}$ in copper and in several copper-base alloys, and for the solution of $\underline{C}$ and $\underline{O}$ in iron and in several types of steel, Fig. 4.2c.

These are the types of gas–metal relationship that are most frequently used. Others may be studied in the texts referred to here and in Chapter 2; the most important of these are the effects of solute or alloying elements on the solubility of gases in metals, gas concentration—temperature relationships when considering protective atmosphere problems, and activity graphs showing the extent of the deviations in solubility from those ideally predicted.

Gas measurement in melting practice. Problems arising in measuring the gas content of metals in melting furnaces differ from equilibria measurements in two important respects. Firstly, the composition of the furnace gases is usually complex, and this is also frequently true of the metal, so that a number of equilibria relationships may be involved, and the departure of binary systems from idealised behaviour may be large. Secondly, the degree of equilibrium reached cannot be readily ascertained for all possible reactions. For these reasons the main approach to gas measurement in metals in foundry processes is by direct determination of the gases dissolved, wherever this is possible, and subsequent reference to the equilibria relationships for the better understanding and control of the process. It is conceivable that in the not too distant future the gas content in molten metals, at least in some melting processes, will be calculated from measured data of the furnace atmospheres. The present practice of direct gas measurement or estimation can be carried out in a number of ways.

Quantitative and analytical methods of gas determination. The most widely used method in this group is to take a sample from the liquid metal in such a way as to retain all the gas in the solidified sample, and analyse this for the gas content. The gases are then extracted from the remelted sample by vacuum fusion, or removed by diffusion at adequately high temperatures from the solid sample; this is known as vacuum solid state diffusion. The amounts of the extracted gas can be inferred by measuring the pressure in an apparatus of known volume before and after separating each component gas from the mixture, Fig. 4.3a.

These are among the most accurate methods of gas analysis in metals, but for foundry purposes slightly lower accuracy is acceptable if the time and costs of estimation can be reduced. Figure. 4.3b shows one such quick method based on direct measurement of gas content in a liquid metal. A small volume (about 1 cm^3) of a neutral or insoluble gas is bubbled through the liquid metal in a closed pumping system, so that the same bubble is passed repeatedly through the melt until equilibrium is reached between the original gas dissolved in the liquid metal and the gas which has diffused into the bubble. The bubble is then passed over a catharometer and, as the thermal conductivity of the bubble gas is sensitive to the amount of gas diffused into the bubble from the liquid metal, the apparatus can be calibrated to read directly the amount of gas dissolved in the metal. This method has been found useful for determining the hydrogen content in some alloys (particularly Al-base) up to about 1,000°C. At higher temperatures it is difficult to obtain materials for the probe which are capable of being immersed in the molten metal without physical or chemical complications.

Semi-quantitative and indirect methods. It is often less important to establish the actual amount of gas dissolved in the metal than to ascertain that the gas level has been brought below a certain 'safe' casting limit. Several

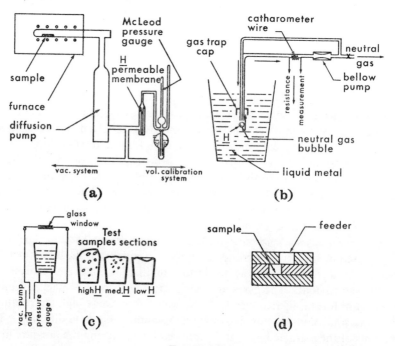

FIGURE 4.3

(a) Solution gas determination in a solid sample by vacuum extraction (e.g., $\underline{H}$ in Al, Cu, Ni or Fe).

(b) Ransley gas bubble method of simple solution gas determination in liquid metals (e.g., $\underline{H}$ in Al or Cu).

(c) Vacuum solidification method of gas content estimation for molecular or reaction gases (Pfeifer test).

(d) Density sample casting mould for estimation of molecular or reaction gases.

such gas control test methods have been developed, some for specific alloys and others for general applications.

The *reduced pressure solidification gas test* (Fig. 4.3c) is frequently used for some non-ferrous alloys. A sample taken from the melt and solidified under reduced pressure is sectioned vertically and examined for evidence of macro- or micro-porosity (or cavities) against a set of 'standardised' specimens of known dissolved gas content.

As the amount of porosity observed in the solidified sample is proportional to the gas content of the melt, a calibration graph for density comparison, or standard porosity samples for visual comparison, can be prepared beforehand. Solidification under reduced pressure increases the tendency of the gas to come out of solution (Sieverts' law). The measurement requires careful control of the time and temperature variables of the test in order to achieve reproducible and reliable results. A modification of this test has been proposed in which the pressure (or degree of vacuum) at which the first bubble forms is measured, but while the principle is theoretically sound the test has been found difficult to apply in practice.

Instead of using a reduced pressure for solidification of the test sample, a test can be made at normal atmospheric pressure, provided the sample (a small cylinder) cools slowly enough during solidification for most of the gas above the solid solubility limit to come out of solution and form cavities within the solid sample. In a test of this type it is necessary to ensure that no cavities are formed in the sample due to volumetric shrinkage, and adequate feeding is critical, Fig. 4.3d. A disadvantage of such a sample is that it requires a subsequent density determination for gas content evaluation, but recent improvements in direct reading balances have made this kind of measurement much simpler. As with the reduced pressure test, the density test can be calibrated in advance.

Quality tests. As gas–metal problems are as old as the casting process itself, one can readily see why a large number of melt quality tests have been empirically developed and are still carried out in the industry with a varied degree of practical success. These depend either on observation of certain phenomena during solidification or on an examination of such properties of cast metals as may be sensitive to the volume of cavities formed by the gas liberated during the solidification. In the former group are tests in which samples cool unidirectionally and upwards, so that the gas liberated can be observed at the top surface, or give a distinct appearance of the top surface rising or falling; level-set tests used for tough pitch copper are of this kind. In the second type of test, fracture, bend, plasticity and fracture after deformation of the sample are examined. The degree and type of discoloration of the fracture due to cavities or the extent of the deformation can be used as an indication of the gas content of the melt. The uses and limitations of such tests can be assessed only for a specific alloy and for a given melting practice and application of the cast metal. The importance and relevance of gas testing in quality control in general is discussed in Chapter 11.

4.3.2 Gas removal from liquid metals

Amongst the metals widely used in industry, tin, lead and zinc do not dissolve gases in sufficient quantities to raise the problem of gas removal in melting practice. On the other hand, aluminium, magnesium, copper, nickel, iron, titanium, and most alloys based on these metals, dissolve one or several gases, or contain elements which may react with one of the dissolved gases. Table 4.1 shows some examples of gas solubilities, typical inclusions occurring in the cast structure due to the gases and treatment. It is clear that hydrogen, oxygen, sulphur and carbon are the most likely examples of gas contamination in industrially cast metals. The various degassing methods are based on the application of different principles in specific cases and will be discussed separately.

TABLE 4.1 *Examples of gas solubilities in various metals, typical gas or solid inclusions that result and methods used to remove the inclusions*

Metal or Alloy Group	Gas	Equilibrium solubility H in $cm^3/100$ g at s.t.p. O, S, C and N in wt%		Typical inclusions due to gases in melting and casting processes	Examples of methods used to control or remove gases or solid inclusions
		Solidus	Liquidus		
Magnesium	H	18	26	H_2	Gas swept out by chlorine gas or decomposition products of chlorides such as C_2Cl_6, C_6CL_6, CCl_4.
	O or N	—	—	MgO or Mg_3N_2	Mg, Ca, Na, K chloride fluxes used to remove oxides. Protective atmosphere of SO_2 used during pouring. Inhibitors H_3BO_3, NH_4HF_2 and sulphur (up to 4%) used in the moulding sand.
Aluminium	H	0·04	0·69	H_2	Degassing, as for magnesium, but nitrogen gas may also be used.
	O	<0·003	—	Al_2O_3	Sodium and potassium chlorides and fluorides used as fluxes.

TABLE 4.1 (*Contd.*)

Copper	H	2·0	5·3	H_2 and H_2O	Degassing by nitrogen or by deliberate oxidation followed by controlled deoxidation.
	O	0·6	0·39 at eutectic, 1,065° C	Cu_2O	Deoxidation by carbon, phosphorus (as copper-phosphorus alloy) or lithium or calcium.
	S	0·0005	0·77 at eutectic, 1,067°C	Cu_2S	Kept low by careful selection of charge materials and control during melting.
Nickel	H	18·5	39·0	H_2 and H_2O	Degassing by deliberate oxidation followed by controlled deoxidation.
	O	0·02	0·236 at eutectic, 1,436°C	CO, CO_2, NiO, or complex oxides in alloys	Deoxidation with carbon, magnesium, silicon, aluminium.
	S	—	21·5 at eutectic, 649°C	NiS or complex sulphides in alloys	Kept low by careful selection of charge materials and control during melting. Some desulphurisation achieved by adding Mg or Ca.
	C	0·55	2·2 at eutectic, 1,318°C	Graphite, CO, CO_2	CO and CO_2 kept low by controlling access to oxygen sources.
Iron	H	6·8	27·0	H_2, H_2O	During refining, CO lowers hydrogen content ('carbon boil').

TABLE 4.1 (*Contd*).

Iron (*Contd.*)	O	<0·003	0·16 at eutectic, 1,523°C	Complex oxides, CO, CO_2	Some prevention by effective slag cover. Deoxidation by Si, Mn, Al, Ti, Zr, Ca, Mg or their combinations.
	S	<0·05	31·5 at eutectic 988°C	Complex sulphides	Reduction in sulphur content by basic slag treatments or metals (Mg, Ce, Ca). Neutralisation by additions (Mn).
	N	0·011	0·040	Complex nitrides	During refining, CO lowers nitrogen content ('carbon boil'). Neutralisasion by additions (Al, Ti).

Removal of hydrogen. The removal of hydrogen from liquid metals is based on the application of Sieverts' law. This law indicates that the solubility of hydrogen is controlled by the alloy composition and temperature (the magnitude of K_t), and by the partial pressure of hydrogen in the furnace atmosphere, usually formed by dissociation of water vapour or other gases. Hydrogen contamination can therefore be completely prevented by using a hydrogen-free atmosphere in melting. However, in industry whenever possible the melting practices used are the least costly consistent with the properties of the castings required; in consequence, for most common metals except titanium, the normal furnace atmospheres lead to hydrogen solution in the melt, and the problem of hydrogen removal has frequently to be solved. All the hydrogen degassing methods used are essentially different technical solutions of applying Sieverts' law.

If a melt containing hydrogen is held sufficiently long at a low pressure (vacuum) and the system is continuously pumped out, then $\underline{H}$ will be reduced to a very low content, due to the low partial pressure of H_2 above the melt. The same applies to other dissolved gases irrespective of their previous origin. In principle this method can also be used for all alloys where evaporation of alloying elements is negligible. In present practice, for the sake of economy, gas removal by vacuum melting and vacuum degassing is generally combined in one process, but vacuum degassing as a separate treatment is also used with some iron, nickel and copper alloys. In the latter case, the metals are melted in a conventional furnace, but subsequently vacuum degassed in a sealed ladle.

There are many instances where either simple molecular or compound gases are insoluble or neutral to a given alloy. If such a gas is bubbled through the liquid metal, and provided it is free from hydrogen or compounds containing hydrogen, then, following Sieverts' law, the $\underline{H}$ dissolved in the alloy will diffuse into the neutral gas bubbles. In other words, $\underline{H}$ in the metal strives to establish an equilibrium with H_2 molecules contained in the neutral gas. As the neutral gas is at the start hydrogen-free, it will absorb some $\underline{H}$ from the liquid metal, and with neutral gas bubbles continuously rising through the liquid metal $\underline{H}$ is continuously removed, and the metal is degassed. The rate of hydrogen removal is controlled by the number and volume of neutral gas bubbles, the length of path and the speed of rising through the melt. Several techniques of introducing the neutral gas can be used as well as several kinds of neutral gas. For example, for aluminium alloys nitrogen or chlorine can be used as gases directly from cylinders; alternatively, C_2Cl_6 (or similar compounds) can be submerged as solid pellets below the surface of the melt, liberating on volatilisation and reaction CCl_4, Cl_2 and $AlCl_3$, which pass as gas bubbles through the melt degassing it in turn. For copper alloys either nitrogen or CO_2 can be used, while for steels the choice of neutral gas is limited to the inert gases such as argon, although in the refining stage of melting steel CO and CO_2 are the main agents for hydrogen removal. Grey or malleable irons are not normally degassed for industrial purposes.

If a melt is allowed to solidify in such a way that $\underline{H}$ continuously concentrates in the liquid due to low $\underline{H}$ solubility in the solid (Fig. 4.2b), and solidification progresses towards the free top surface, then the melt will be partly degassed. The amount of residual $\underline{H}$ in the completely solidified metal will be determined by the maximum solid solubility, and there should be no gas porosity. The application of this principle of degassing is referred to as a presolidification degassing process. Implicitly this process requires subsequent remelting of the metal for it to be poured into moulds, and in practice it is used to only a very limited extent.

Some alloying elements may raise, others reduce, the solubility of hydrogen (or other gases) in metals. In the latter case the problem of degassing the solvent metal may be reduced or eliminated. For example, the high vapour pressure of zinc in brasses makes it often unnecessary to degas copper in making these alloys, whereas for many other copper base alloys hydrogen removal is generally necessary.

Removal of oxygen or deoxidation. The importance of oxygen solubility in metals differs from that of hydrogen mainly on account of the greater tendency of oxygen to form stable compounds with metals. If such compounds are insoluble in liquid metals at the normal melting temperatures, then deoxidation is unnecessary as indicated previously for aluminium and magnesium. The same principle holds, for example, for tin, lead, cadmium and zinc. Flux and slag treatments of surface oxides with these metals has been considered in Chapter 2.

For metals that dissolve oxygen, such as copper, nickel and iron, the solubility of oxygen in relation to the furnace atmosphere can be treated like hydrogen in terms of Sieverts' law. On the other hand, the application of this law to the removal of oxygen from solution is limited in practice; vacuum and neutral atmosphere degassing are possible by virtue of the low oxygen pressure above the melt, but the rate of diffusion of $\underline{O}$ is too low for neutral gas bubbling or solidification degassing to be practicable. Deoxidation of alloys is therefore effected by applying the principle of the relative stability of oxides, which was discussed in Chapter 2 in relation to free energy diagrams of metal oxides. To a liquid metal solvent M_a, containing oxygen in solution, a metal solute M_b is added, the condition being that oxide M_bO is more stable than the oxide M_aO. Metal M_b is a satisfactory deoxidant for metal M_a if the following additional conditions obtain: the product of deoxidation, the oxide M_bO, should separate readily from the liquid metal; the properties of the metal M_a should not be detrimentally affected by any residual amount of M_b in solution; and finally the amount of residual oxygen in solution should not have a significant effect on the properties of the cast alloy.

The deoxidation reaction can be represented by the equation

$$\underline{M_b} + \underline{O} \rightleftharpoons M_bO$$

where $\underline{M_b}$ and $\underline{O}$ are in solution in M_a, and M_bO is a gaseous liquid or solid oxide. In principle the equilibrium constant K of this reaction can be determined from free energy data, provided the active masses, or activities, of $\underline{M_b}$ and $\underline{O}$ in a particular solvent M_a have been previously established experimentally or by calculation. In practice, for lack of numerical data and because of the non-ideal behaviour of many systems, the value of K is determined experimentally. It is also necessary to consider the kinetics of mixing M_b in M_a, and particularly the separation of M_bO from the melt.

The free energy diagrams show many industrially used metals in the upper and a wide choice of possible deoxidants in the lower part. The actual choice of deoxidant depends therefore on the additional requirements already discussed. For example, phosphorus is the most generally used deoxidant for several copper alloys, and the liquid deoxidation product separates readily. But for high conductivity copper, where residual phosphorus would impair this special property, melting is carried out under a neutral atmosphere, or lithium is used as a deoxidant. For steel a variety of deoxidants is used, and generally several at the same time; silicon, manganese and aluminium are the most common, but in addition titanium, calcium, zirconium and others are often used. Methods, quantities and other metallurgical or practical aspects of deoxidation are further discussed in the suggestions for reading at the end of the chapter. The particular problem of inclusions in cast structure arising from failure of deoxidation products to separate completely from the melt is discussed in 4.4.4 and in Chapters 7 and 11.

Removal of sulphur, carbon and nitrogen. Control of the sulphur, carbon and nitrogen content of metals may be effected by vacuum or neutral atmosphere melting, but when using a normal furnace atmosphere these elements cannot always be satisfactorily removed from the melt in the same way as hydrogen and oxygen. Sulphur behaves similarly to oxygen, but there is not such a wide choice of stable sulphide-forming elements whose performance can be predicted directly from free energy graphs. Some suitable elements are however available, and sulphur is removed from grey irons by treatment with magnesium, cerium or calcium, but the bulk of the sulphur is generally removed by applying slag metal reactions in a preliminary refining process. The same is true if carbon and nitrogen in an alloy have to be removed for some specific purpose, and there is no simple technique for substantially complete removal of these two elements in normal foundry melting. Sulphur, carbon and nitrogen in non-ferrous and general grey iron melting practice are therefore controlled, not by degassing, but by using charges of correct composition at the start, and by controlling the melting conditions to avoid possible contamination of the liquid metal; while in bulk steel melting both sulpur and carbon are first brought down to the desired level by means of oxygen blowing or using deoxidising slag, and the metal is subsequently further deoxidised just before or during the pouring. In some alloys the harmful effects of certain elements arising from gaseous reactions can be controlled by the addition of suitable elements to 'neutralise' the gas effects. For example, manganese is added to steels and cast irons to form stable sulphides (inclusions being less harmful than dissolved S) and aluminium or titanium can be used in some cases to control the behaviour of nitrogen.

4.4 GASES AND THE STRUCTURE OF CAST METALS

The complete removal of gases from liquid metals is restricted by physical and economic factors. Physical restrictions derive from the fact that gas elimination methods are largely based on the application of the laws of equilibria, and these postulate clearly that many chemical reactions, such as degassing reactions, seldom proceed to completion in one direction. For these reasons it is important to consider metallurgical problems arising from the presence of residual gases in liquid metals. Such problems fall into two groups: the behaviour of gases during the solidification and the state of residual gases in the structure of solid metals.

4.4.1 Gases and solidification

As a metal begins to solidify, the behaviour of a dissolved gas will depend on its concentration. If the amount of gas dissolved is below the solid solubility value (Fig. 4.2b), then the gas remains in the solid state as a solute element. With rates of cooling more rapid than the equilibrium rate, gas in excess of the solid solubility limit can be retained in the solid metal, thus giving a supersaturated solution. This dissolved gas behaves analogously to any other solute alloying element. The gas solution may be interstitial or substitutional and gas distribution may be homogeneous or heterogeneous, preferred or random.

From the point of view of casting metallurgy, the problems which arise when the amount of gas in the liquid metal exceeds that which can be retained in solution in the solid state are of particular interest. The concentration of the gas in the remaining liquid will increase as solidification progresses, and at a certain stage gas bubbles are nucleated and grow in the following manner:

$$\underline{H} \rightleftarrows H_{2(g)}$$

or

$$\underline{H} + \underline{O} \rightleftarrows H_2O_{(g)}$$

The former is an example of the behaviour of a simple molecular gas, and the latter that of the reaction which may occur when two dissolved gases are present. The gas bubbles formed may rise to the free surface and escape, or more generally be trapped in the solidifying alloy forming cavities (voids or porosity) of different size, shape and distribution. Nucleation and kinetics of gas bubble formation during solidification have not been studied in great detail, but certain general features have been established. The nucleation of gas bubbles is of a heterogeneous character, and bubbles generally form on the solid matter present, such as growing crystals, suspended impurities or the walls of the mould. The pressure of gas at which growth of the bubbles is possible is controlled by the following general condition:

$$p_g = p_a + p_h + \frac{2\sigma}{r},$$

where p_g is the internal pressure of the gas as defined by the equilibrium constant, p_a is the atmospheric pressure, p_h is the hydrostatic pressure, σ the metal surface tension around the bubble, and r the bubble radius. The above relation explains the requirement for heterogeneous bubble nucleation, as for very small values of r the bubble pressures required for growth would be very large. It is also important to emphasize that, for gas liberation on solidification, it is the equilibrium between the internal and external pressure that matters, and not the partial pressure of the dissolved gas in the atmosphere.

4.4.2 Gases in solid metals

In the metallurgy of cast metals the problems associated with gas voids or cavities are more frequently encountered than the physical problems arising from dissolved gases in solid metals. These are concerned with the size, the shape, and the macro- and micro-distribution of cavities, Fig. 4.4. These characteristics are related to the concentration and nature of the gas or gases,

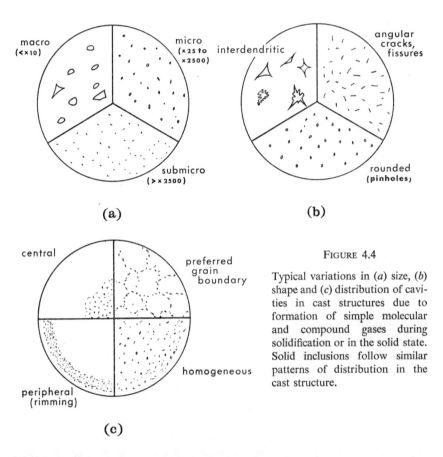

(a)

(b)

(c)

FIGURE 4.4

Typical variations in (a) size, (b) shape and (c) distribution of cavities in cast structures due to formation of simple molecular and compound gases during solidification or in the solid state. Solid inclusions follow similar patterns of distribution in the cast structure.

the pattern of solidification, the volume contraction phenomena occurring during solidification, and the external conditions of bubble formation. Cavities will tend to be spherical when forming largely in contact with the liquid alloy but more angular if they form during the later stages of solidification, a particularly common occurrence when they are formed by a reaction gas. For example the amount of dissolved oxygen left in an alloy after deoxidation is very small, but it may concentrate sufficiently in the residual liquid to react, e.g. with hydrogen or carbon, and give rise to gaseous products towards the end of the freezing process. In some cases cavities may originate due to the volumetric contraction on freezing and they then act as free space volumes into which the gas diffuses when it exceeds the solubility limit. Both the macro- and micro-pattern of gas cavity distribution are related to the type of crystals forming, and the direction of solidification. For example, they may form between the arms of dendritic crystals, but when formed at the end of solidification they are mainly located at the boundaries between crystals (Chapters 5 and 6).

4.4.3   Gases arising in the course of mould filling

In addition to the gases dissolved in alloys during melting, further 'gassing' of castings can occur during the mould filling stage of the casting process. The main sources are (a) physical occlusion arising in flow (page 228), (b) physical occlusion of mould or core gases due to the lack of permeability or venting of moulds or cores (8.8.1) and (c) chemical reactions at the metal-mould interface (8.10.1). The location, shape and surface appearance may help in differentiating gas cavities due to these sources from those due to the dissolved gases in the melt. In general, special experimental techniques have to be used for such identification, and in some cases both the composition and the physical character of cavities from either source are identical.

4.4.4   Gases and the properties of cast metals

The practical objective of removing gases from metals can best be appreciated when the properties of cast metals are considered. For statuary castings, the metallurgical quality of the casting is judged mainly by its surface perfection and the soundness of the casting skin, whereas optimum soundness and freedom from gases are required in castings used for jet engines. Gas–metal problems in practice are therefore strongly influenced by the effect of gases on the properties required in the cast metal.

Dissolved gases. The properties of cast metals most strongly affected by dissolved gases are those related to plasticity and fracture of metals under stress (Chapter 7). The dissolved gases lower the ductility of metals and reduce the value of the external stress required to initiate the fracture. The magnitude of these effects varies with different alloys and is generally small, but for the particular borderline condition of transition of a gas from the dissolved state to that of submicroscopic or microscopic porosity the effect may be large. Furthermore, a dissolved gas in the supersaturated state tends to separate into the molecular state as the conditions of equilibrium demand. Gas separation in the solid may in this way lead to the generation of porosity (as in annealing) or of internal stress patterns (in the elastic range at lower temperatures) in the surrounding crystals, and even to the formation of cracks of finite dimensions, such as hydrogen cracks or flakes in some cast alloy steels. For the majority of casting alloys and casting applications, the property effects of dissolved gases below the solubility limit are, however, of secondary importance.

Cavities. The problem of gas cavities is extremely complex. Broadly speaking they may be discussed under the headings of cavities in cast metals which are subsequently worked (ingots), and cavities in shaped castings.

Cavities in ingots may cause fracture during plastic deformation, or they may weld up so that the ingot partly or fully regains the properties of sound metal. The danger of deformation failure increases with alloys of high mechanical strength, while softer and more plastic alloys, a good example of which is rimming steel, may contain a large volume of such cavities without detriment to either working and fabrication processes or the properties of the final product. Full welding of gas cavities during working depends also on the condition of their surfaces; for example, an oxide film may prevent the full joining of the opposite surfaces of a cavity.

Gas cavities in shaped castings are however permanent and their effect is therefore estimated entirely from the point of view of the subsequent application of the castings as, for instance, mechanically stressed components, containers for liquids or gases at various pressures, and unstressed components. In the first two applications gas cavities lower the performance of the casting; acceptance standards vary from case to case, and this question is discussed in Chapter 11. For unstressed components the effect of gas cavities is judged entirely on the functional requirements of particular castings. In a few cases gases are intentionally added to liquid metals, because cavities formed may help in overcoming some other more directly harmful production problems. For example, in gravity die casting of some light alloys, a limited and controlled volume of gas cavities may be used to counteract an unfavourable distribution of shrinkage cavities (Chapter 10).

Inclusions originating from gases. Impurities. Some products originating from gases appear in the cast structure as solid phases such as oxides, sulphides, carbides and nitrides (endogenous inclusions). These products form either in the liquid metal or during solidification (e.g., as a result of deoxidation) and, like gas cavities, they vary in size, shape and distribution. Solid inclusions of this type have effects similar to cavities (see Fig. 4.4); they act essentially as discontinuities and therefore affect the plastic behaviour and stress to failure of alloys, although in some alloys they can be used for structure strengthening. Structure and structural defects are more fully discussed in Chapter 7.

Other inclusions, in many respects micro- or macroscopically similar to the endogenous inclusions, find their way into the cast structure from outside sources (refractories, slag, flux, mould, mould dressing, etc.). These are commonly described as exogenous inclusions.

Inclusions are often referred to as impurities. The term impurity is, however, generally used to describe the presence in the alloy of any solute elements or of any separate constituents which are not intentionally added to it. The importance of impurities in the formation and on the properties of cast structure are discussed in Chapters 5 and 6.

4.5 SUMMARY

Gas–metal problems are of very wide interest in casting metallurgy. Solubility laws can be applied to deal with the problems of occurrence and removal of gases. However, in most cases a quantitative treatment for the purpose of process control is feasible for simple binary solutions only. This is partly due to the complexity of gas–alloy systems in industrial applications and partly to their non-ideal behaviour and the consequent need for more experimental data. For these reasons many of the gas–metal control problems in foundries are treated at present by direct and *ad hoc* experimental methods.

FURTHER READING

SMITHELLS, C. J., *Gases and Metals*, Chapman & Hall, London, 1937.
SMIALOWSKI, M., *Hydrogen in Steel*, Pergamon, Oxford, 1962.
EASTWOOD, L. M., *Gases in Light Alloys*, Wiley, New York and London, 1946.
Gases in Metals, Symposium of the American Society of Metals, 1953.
Gases in Metals, Symposium of the Australian Institute of Metals, 1954.
SIEVERTS, A., Die Aufnahme der Gase durch Metalle, *Z. Metallkunde*, 1929, **21**, 37.
ALLEN, N. P. and HEWITT, T., The equilibrium of the reaction between steam and molten copper, *J.Inst.Metals*, 1933, **51**, 257.
RANSLEY, C. E. and TALBOT, D. E. J., Wasserstoff Porosität in Metallen, *Z. Metallkunde*, 1955, **46**, 328.
PHILLIPS, A. J., Separation of gases from metals, *Metals Technol.*, 1947, T.P. 2208.
DARDEL, Y., Purification of Al and its alloys, *Metals Technol.*, 1948 T.P. 2484.
Non-ferrous melt quality tests, *Br. Foundryman*, 1960, **53**, 120.
JAMES, J. A., Techniques of gas analyses in metals, *Metall. Rev.*, 1964, **9**, 93.
SIMS, C. E., The non-metallic constituents of steel, *Trans. A.I.M.E.*, 1959, **215**, 367.

CHAPTER 5

SOLIDIFICATION: CRYSTALLISATION OF METALS AND ALLOYS

5.1 INTRODUCTION

Metallurgically the solidification of a liquid metal in a mould is the focal point of the whole foundry process. It is the shortest of the several events involved in the process, and yet it is the most vital one. The initial metallurgical problems of preparing the charge, melting and controlling the liquid metal, and the subsequent ones of examination and evaluation of the properties of castings, are all directly related to the objectives and the outcome of the solidification process. The major objective of metallurgical control of solidification is simply to achieve the required structure in a cast metal. In order to discuss fully the term 'required structure', it is necessary to consider three distinct aspects:

(a) Crystallisation of the liquid metal (liquid–solid phase transformation), with special reference to the origin and characteristics of the solid crystals formed.

(b) The effects of non-equilibrium on crystallisation, and of physical changes occurring during or immediately after crystallisation that have a bearing on the properties of the crystals and of the general structure.

(c) The properties of cast metals in relation to (a) and (b), taking account of the effects of the size and shape of castings, and various types of defect which may occur in the cast structure.

These three distinct aspects of cast structure are dealt with in Chapters 5, 6 and 7 respectively.

Crystallisation is a complex field involving various aspects of physical chemistry, thermodynamics and crystallography. The liquid state of mobile and disordered atomic arrangements changes into a solid where the centres of atoms form regular patterns, or crystal lattices, obeying certain rules of crystal geometry. In solid crystals the position of an atom vibrating around its centre is relatively fixed with respect to its neighbours. The position does not vary more than the interatomic distance of the vibrational centres of the neighbours in the lattice, unless thermal or mechanical energy is applied (for example in diffusion or plastic deformation).

79

The liquid or mobile state transforms into a solid and regular lattice arrangement in the following way. A small number of the atoms in the liquid form a minute embryo or nucleus of the solid crystal, which then grows until the liquid is completely used up. In principle a whole body of liquid of a pure metal could be induced to grow into a single crystal, or alternatively the number of nuclei at the onset of crystallisation might be extremely large, so that the final cast structure would be made up of a very large number of very small crystals with differing spatial orientation of their main lattice axes. In practice single crystals are usually only grown under laboratory conditions for special purposes, and these are generally small. Industrial cast metals or alloys invariably contain a large number of crystals or grains of various phases in the structure. The problems relating to the size, shape and arrangement of these crystals with cast structure of a metal or alloy will be discussed in more detail in this chapter.

A number of additional problems arise in this field which have to be considered separately. More important amongst these are the relationships between the composition of the liquid alloy and its crystallisation, and the possible extraneous effects of physical factors such as vibration or other forms of energy. This chapter deals further with the application of the general concepts of crystallisation to explain the structure of industrial cast metals and alloys.

5.2 LIQUID–SOLID TRANSFORMATION (CRYSTALLISATION)

The crystallisation of liquid metals follows the pattern of crystallisation of most other liquids. In fact our early understanding of the mechanism of the solidification of metals is due to the classical work of Tammann on the crystallisation of certain organic liquids. Crystallisation is essentially a phase transformation, the liquid phase changing into a solid as the substance cools, and it may therefore be conveniently studied by considering the energy changes in the process.

On cooling, a liquid loses some of its energy in the form of heat and becomes structurally unstable in relation to the solid state. The main characteristic of liquids is the disorderly, mobile configuration of atoms. When the temperature is lowered, the average distance between the atoms decreases. As the thermal energy which maintains the vibration of the atoms and keeps them apart is lost to the surroundings, the attractive forces between neighbouring atoms become more prominent. On further cooling the attractive forces between atoms may prevent them from moving away from one another any longer, and the liquid gradually becomes fully frozen or solid. Such a process, where the liquid solidifies while retaining largely its previous structure in the solid state (supercooled liquids), is very rare among the methods of liquid to solid state transformation, but occurs, for example, with glass.

Another possible freezing mechanism is for the liquid to transform instantaneously and in one unit from the disordered to the ordered state as a block transformation at a constant temperature. This kind of transformation, like the previous one, can occur when small energy changes are involved, but is infrequent among phase transformations in metals. An example is when one kind of solid phase changes into a crystallographically similar one, such as the austenite–martensite transformation in steels, 6.9.1.

This leaves the third possibility, namely, when only a small fraction of the liquid transforms into a solid nucleus at the start. In this case the formation of the solid nucleus requires a local increase in free energy, but the total energy change, that of the transformed liquid and of the solid nuclei formed, is lowered, thus satisfying the energy requirements of the process to proceed. In this way the liquid transforms gradually into the solid, first by the formation of crystal nuclei, then by their subsequent growth. The energy transformation requirements of this process allow it to proceed at constant or changing temperatures, depending on the composition of the original liquid and of the solid formed. This picture of liquid to solid transition explains in general terms the bulk microscopic or macroscopic nature of the transformation.

5.3 HOMOGENEOUS AND HETEROGENEOUS NUCLEATION

A pure liquid, for example a metal containing one kind of atom only, will on cooling reach a temperature where the closeness or density of packing of the atoms is such that a small number of them may be able to join together and form a solid nucleus. Such a nucleus, made up of the same atoms as the liquid itself, is called a homogeneous nucleus. The energy conditions required for the formation of such a nucleus can be analysed with reference to Fig. 5.1.

The formation of a small nucleus, for this purpose assumed to be a cube with side dimension a, involves a change in energy per unit volume, $\Delta G_{LS}'$. This is energy of freezing

$$\Delta G_{LS} = a^3 \cdot \Delta G_{LS}'$$

The total energy of the system is reduced during solidification by the above amount, and this is plotted in the negative direction in the energy balance in Fig. 5.1a. The formation of a solid nucleus also involves the formation of a new surface, and this interfacial energy per unit surface represents an increase in the energy change of the system, and is represented by a positive plot in Fig. 5.1a. For the cube this is equal to $6a^2 \cdot \Delta Gi'$. The total free energy change is therefore

$$\Delta G = -a^3 \cdot \Delta G_{LS}' + 6a^2 \cdot \Delta Gi'.$$

It can readily be shown from Fig. 5.1*a* that there is a critical size a_{cr} for a stable nucleus. Below this size nuclei redissolve in the liquid, and above it they are capable of growth, because this is accompanied by a total decreases of available or free energy in the system. In other words, a decrease in free energy due to the phase change (L → S) is greater than the increase due to

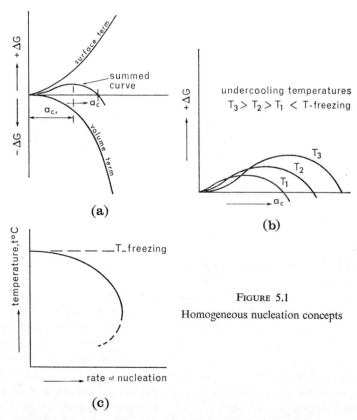

(a)

(b)

(c)

FIGURE 5.1

Homogeneous nucleation concepts

(*a*) Critical size a_{ct} of the nucleus for thermodynamic stability.
(*b*) The effect of undercooling temperature on the critical size.
(*c*) The effect of undercooling temperature on the nucleation rate.

surface boundary formation. These energy considerations explain why the liquid has to undercool below its equilibrium transformation temperature before enough transformation can occur to satisfy the condition of a decrease in the total energy. It is also clear that the greater the undercooling temperature, the smaller the critical size which still satisfies the condition of a decrease in free energy, Fig. 5.1*b*. These relationships can be translated into the dependence of the rate of nucleation on the undercooling temperature as shown in Fig. 5.1c. The rate of nucleation at first increases with the under-cooling temperature when the smaller nuclei are capable of growth, but is later reduced, due to the statistical decrease in the number of collisions of atoms necessary for the formation of nuclei.

These theoretical ideas, though helpful for an understanding of the broad picture of crystallisation, do not unfortunately lend themselves readily to either experimental quantitative interpretation or to direct application in metallurgy. The work of Holomon and Turnbull, in particular, confirmed the basic soundness of such theories, but their work and that of others also clearly demonstrated that in the crystallisation of metals homogeneous nuclei play little, if any, part. The main reason for this is that metals are essentially impure substances. In the normal process of crystallisation, solid metallic or non-metallic particles, some of which could be generally described as impurities or inclusions, are distributed in the liquid itself, and liquid metal is also in contact with the solid surfaces of the mould. When energy con-ditions permit, crystallisation can start on either the container walls or the impurities. This is known as heterogeneous nucleation. The nature, amount and distribution of all heterogeneous particles in liquid metals are difficult to control in foundry practice, and frequently control is not even attempted. Moulds, for example, are chosen for reasons unconnected with surface nu-cleation, while the purification of metals to remove all sources of hetero-geneous nucleation is not generally practicable. Though the main concepts of heterogeneous nucleation are generally accepted, much remains to be-done before it will be possible either to understand it completely or to apply the theory to control fully the structure of cast metals.

The existence of an energy barrier in the formation of homogeneous nuclei has already been explained; this is the main reason why, if solid particles are already available, a liquid will begin to crystallise on them at the equilibrium temperature without the necessity for undercooling. A liquid at the equilibrium freezing temperature can, therefore, be most readily nucleated by fine solid particles of the same composition as the liquid metal itself. Such an idea of artificial homogeneous nucleation is frequently used in laboratory work for initiating the growth of single crystals. Such methods cannot be applied in foundry practice as the pouring temperature of the liquid metal is high enough for solid particles of the same alloy to be fully melted and dissolved.

Heterogeneous nuclei are by definition physically and chemically different from the crystallising liquid. One of the main problems in applying hetero-geneous nucleation theory is to establish the essential properties of such nucleating particles, so that they can be successfully introduced into alloys where solid nuclei, already accidentally present, may not be available in adequate number or kind.

It is readily apparent that surface bonding and crystalline similarity are important factors when considering the behaviour of heterogeneous nuclei. Figure 5.2a shows that a heterogeneous particle may be wetted by a liquid metal or not, depending on the nature of the interface forces between them. The relationship between the surface forces at the commencement of crystal growth at the surface of a heterogeneous nucleus can be represented by the relation

$$\sigma_{HL} = \sigma_{HS} - \sigma_{SL} \cos \psi$$

Clearly the smaller the value of ψ, the greater will be the tendency of the liquid metal to wet the surface of the heterogeneous particle, and the greater will be the potential of this particle to initiate crystallisation. The surface energy characteristics of the numerous particles and impurities commonly encountered in liquid metals are little understood in detail; a good deal of

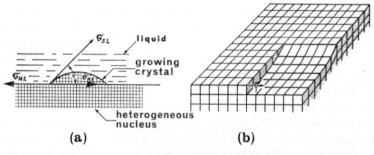

(a)   (b)

FIGURE 5.2

(a) Surface energy relations affecting the wetting of heterogeneous nuclei by the liquid metal.

(b) Screw dislocation at the surface of a growing crystal, showing the next rows to be filled. When the whole crystal face has thus been filled, the screw dislocation is repeated at the next face, one interatomic distance up.

attention has therefore been paid by investigators to the crystalline nature of such particles, which can more readily be studied experimentally. Thus it has been shown that a certain measure of crystalline similarity between the liquid metal and the heterogeneous particles is necessary for effective nucleation, and that a difference in the basic lattice unit size of the two crystal systems should not be more than 15% to 20% for the particles to act as heterogeneous nuclei. However, some particles that satisfy these structural conditions still have no nucleating effect, indicating that crystalline similarity is not the only condition for heterogeneous nucleation to occur. In spite of the limitations in current knowledge of the details of heterogeneous nucleation, it is possible to apply the main concepts outlined above to influence the solidification behaviour of many casting alloys. Such processes are known as grain refinement or inoculation treatments (6.6.1 and Table 6.1). As a rule

PLATE 1

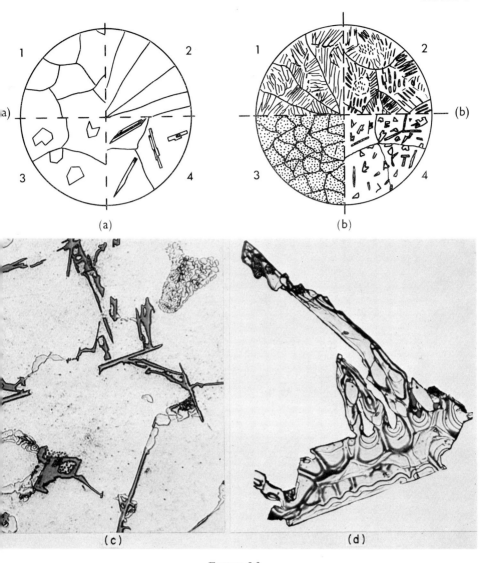

(a)

(b)

(c)

(d)

FIGURE 5.3

(a) Cross-sections of grains or crystals: (1) equiaxed and (2) columnar crystals of pure metals and solid solution alloys, (3) and (4) more regular external surfaces of intermetallic compounds in solid solution matrix, equiaxed (3) and plates or needles (4).

(b) Cross-sections of eutectic structures: (1) lamellar, columnar matrix and growth cells, (2) rod, coarse equiaxed matrix and growth cells, (3) globular, fine equiaxed matrix, (4) anomalous, coarse equiaxed matrix.

(c) Microstructure showing typical intermetallic phases of an Al–base alloy: light $CuAl_2$, dark (SiFeAl) phase. Matrix Al solution. × 260.

(d) Crystal of silicon electrolytically removed from the matrix of an Al–Si eutectic alloy.

PLATE 2

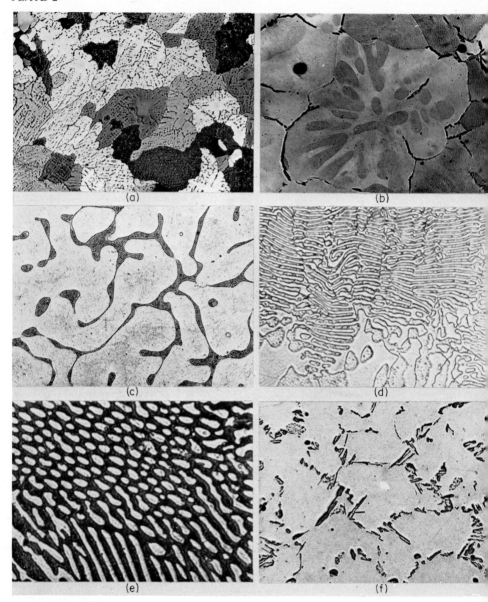

FIGURE 5.7

(a) Microstructure of commercially pure aluminium. Impurities outline the matrix growth (Al–Si–Fe compound). Electrolytic etch, × 38.

(b) Cored distribution of iron and silicon in aluminium matrix, and Al–Si–Fe compound a separate phase. Electrolytic etch, × 38.

(c) Aluminium–10% copper alloy: matrix solution crystals (α–Al), dispersed eutectic of α–Al and CuAl$_2$ (α–Al light and CuAl$_2$ dark). Etched, 0·5% HF, × 45.

(d) Lamellar eutectic of aluminium with 32·7% copper alloy. Light, continuous α–Al, darker dispersed CuAl$_2$ partial divorcement produced by 0·84% zinc. Etched 0·5 HF, × 750.

(e) Rod eutectic antimony and 21% zinc. Matrix almost pure antimony, rods SbZn compound, boundary between two grains shown. Etch 2% FeCl$_3$, × 750.

(f) Aluminium–5% silicon alloy: α–Al matrix and divorced eutectic. A peritectic structure can be similar. Unetched, × 45.

practical application of such processes necessitates empirical tests to ensure the availability and distribution of finely dispersed particles in the liquid metal at its freezing temperature, in order to achieve optimum nucleation effects.

5.4 CRYSTAL GROWTH

The process of nucleation represents a dynamic and instantaneous change at the required temperature and generally it is difficult to study experimentally owing to the opaque nature of metallic liquids. The rate of crystal growth, on the other hand, can be controlled, and at a given stage it can be arrested or made to proceed in a desired manner. Such techniques allow the growing surface of metallic crystals to be examined, making it possible to obtain a reasonably clear picture of the main characteristics of their growth during crystallization.

The arrangement of atoms in a crystal is far from perfect, and it contains a large number of lattice defects described as point, line and plane defects (Chapter 7). Crystal growth on the atomic scale is believed to be initiated at some of such imperfections and proceeds by atoms becoming attached across the planes of the solid nucleus, until a plane is fully covered. When a plane has thus grown thicker by one atomic layer, the fault which has initiated the growth is transferred automatically to the new plane, which grows in the same way, Fig. 5.2b. Linear propagation of atomic attachments achieved in this manner appears to be more likely than instantaneous freezing of a monoatomic layer on the available surface of the solid already present.

The microscopic and macroscopic features of the growth of metallic crystals are a good deal easier to study than the processes on the atomic scale described above. It is therefore in this field that most of the progress in studies of the growth forms of metallic crystals has been made. In discussing the present position, it is convenient first to consider the main types of metal crystals observed in cast structure, and then to analyse the manner in which such crystals could have grown.

5.4.1 Macroscopic morphology of metal crystal growth

With mineral crystals, the regularities of the external surfaces give some indication of the essential crystalline properties of the mineral, The type of crystal system, the axes of symmetry and other macroscopically observed features are all derived from the ordered arrangement of atoms in the crystal. With metals, however, the conditions of crystal growth are usually such that the external surfaces of the crystals are irregular. Their internal crystalline nature is however identical to that of minerals and is usually examined by analysing the patterns obtained by the reflection of waves of suitable short wavelength, such as X-rays, passed through the crystal. The particular arrangement of the atoms in various kinds of crystals can be deduced from such reflection patterns.

Typical examples of the shapes of metallic crystals are shown in Fig. 5.3a, Plate 1. These macroscopic crystals can sometimes be clearly seen with the naked eye or at small magnifications (up to × 10), and are referred to as macrocrystals or macrograins. The structure of a casting revealing such crystals after suitable metallographic preparation is known as the macrostructure. Smaller crystals are described as microcrystals, and the structure as the microstructure, Fig. 5.3c. In castings both macro and micro size crystals are encountered, and by controlling the nucleation and growth process the size of crystals can be varied within fairly wide limits (Chapter 6).

The boundary surfaces of the crystals in the cast structure are usually quite irregular, particularly with pure metals and solid solution alloys. The reasons for this are discussed in 5.4.2. When such crystals are roughly spherical in shape, they are referred to as equiaxed, and when elongated in one direction they are described as columnar. Intermetallic compounds, which often obtain in alloys of two or more metals, have certain characteristic growth properties, and as a rule these crystals display flat surfaces, which in some cases resemble mineral crystals. In general, however, their crystal shape is too irregular for symmetry studies from surface measurements only, Fig. 5.3d. Intermetallic compound crystals may also be classed as equiaxed or columnar, but they usually reveal a far greater variety of shape than solid solution crystals and form, for example, polyhedra, plates and needles. While the crystals shown in Fig. 5.3 are typical, pure metal and solid solution crystals can be made with flat or plane growth surfaces, and some intermetallic compounds also grow with irregular external surfaces. Characteristic features of the shape of crystals in the cast structure are generally studied by polishing and etching techniques which are described in textbooks of metallography.

The reasons for the variations in the external shape of crystals, as shown in Fig. 5.3, require some consideration. It can be shown experimentally that equiaxed crystals tend to form when the heat of fusion liberated during crystallisation disperses evenly in all directions, while columnar crystals form when the heat flow is unidirectional, the crystals usually growing inwards from the mould wall. However, elongated intermetallic phase crystals can be obtained with a multi-directional heat flow. These two observations indicate two important macroscopic characteristics of metal crystal growth, the sensitivity to heat flow, and the dependence on the nature of the bonding forces of the atoms in the crystals and hence the surface energies of the various planes of growth.

5.4.2 Microscopic morphology of metal crystal growth

The small scale features or microscopic morphology of growth are of special interest as they may account for the origin of the observed types of macroscopic crystals. Microscopic morphology has been studied by a number of different experimental techniques. The classical metallographic method is to cut across a cast sample and then polish and etch the surface. In recent years special solidification techniques have been developed, the most fruitful of which is due to Chalmers. In this method solidification is made to proceed unidirectionally, and at a given stage of crystal growth the residual liquid is decanted and the growing crystal surface or interface between solid and liquid is examined. By studies of this kind it has been shown that the microscopic morphology of metal crystal growth can be classified into three main types, plate (plane), cell and dendritic, Fig. 5.4. When different growth forms or other variations in the structure of a crystal are revealed within a single macro- or micrograin, these are often referred to as substructures. Most of such features occur on a microscopic or even finer scale. The smallest scale imperfections are those occurring on the atomic scale, and these are revealed by special techniques of electron microscopy.

The plate (platelet or plane) type of crystal growth is explained by the fact that certain planes of the basic crystalline pattern can grow faster than others. The plane crystal growth interface is characteristic of all crystals, but is more marked in the intermetallic compounds than in pure metals or solid solution alloys. The planes as a rule are not perfectly smooth but display a number of irregularities. The pronounced direction of growth is perpendicular to the plane in pure metals and solid solution alloys, but in intermetallic compounds it is often parallel to the growth plane.

A cellular form of crystal appears to develop from the plane type: valleys or corrugations resembling an arrangement of cells of mainly hexagonal arrangement form on the crystal growing faces. It has been shown that metals of extremely high purity tend to exhibit the plane type of growth, and that either small additions of solute atoms or a more rapid rate of cooling may induce the cellular type of growth. The transition from plane to cellular microscopic morphology thus depends on the metal composition and a departure from equilibrium cooling conditions, i.e. on the growth kinetics (solute segregation), Chapter 6.

The dendritic type of crystal growth, Fig. 5.4c is so frequently encountered under normal conditions of crystallisation of metals and alloys that not only was it the first to be observed, but until relatively recently it was thought to be the only type of metal crystal growth morphology on both the microscopic and the macroscopic scale. Recent studies have shown, however, that dendrites can be looked upon as one of the possible changes of form in the growth of the crystals and occur when the cellular type of growth becomes unstable, for example, by further increasing either the solute content or the rate of heat abstraction during crystallisation. The dendritic growth is in general an outcome of the fact that the growing crystal is trying to adjust itself to the requirements of the changing compositions during freezing (solute concentration gradients at the interface), heat of fusion flow conditions (surface and volume ratios) and the preferential growth habit of different crystal planes (surface energies involved).

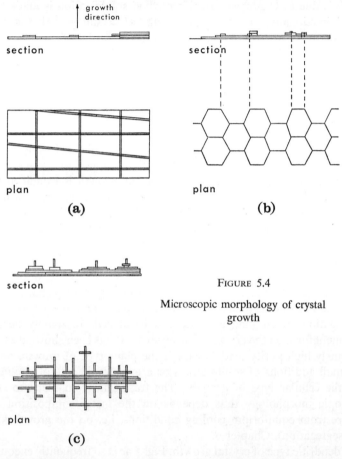

FIGURE 5.4

Microscopic morphology of crystal growth

(a) Microscopically flat crystal planes with submicroscopic imperfections (steps, striations).
(b) Microscopic cellular pattern on the crystal faces; cell planes are not atomically smooth (cell pattern due to grooves or ridges).
(c) Branching, skeleton or dendritic growth ranging from microscopic to macroscopic dimensions, microscopic faces of the skeleton having growth characteristics as (a) or (b).

Micromorphological studies of metal crystal growth have thus shown that three major parameters can be used to analyse the problems of growth: the alloy composition (including the type of crystal system), the rate of crystal growth and the prevailing temperature gradient. The effects of these and other factors on crystal growth and the cast structure in general are discussed in Chapter 6. The micromorphology of crystal growth thus explains the origin of irregular outer surfaces of macrocrystals. This irregularity is further emphasised by the fact that crystals growing from various centres eventually meet and their surfaces of contact, or grain boundaries, are irregular because of the different orientation of the crystals in space.

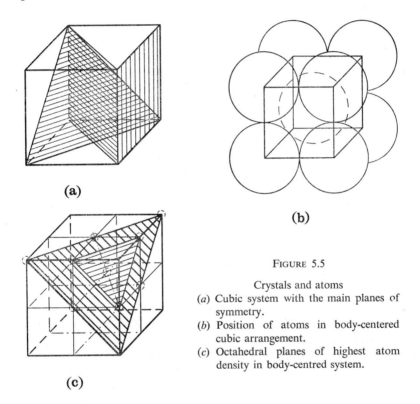

(a)

(b)

(c)

FIGURE 5.5

Crystals and atoms
(a) Cubic system with the main planes of symmetry.
(b) Position of atoms in body-centered cubic arrangement.
(c) Octahedral planes of highest atom density in body-centred system.

5.4.3   Lattice characteristics of metal crystals

The atoms in a crystal can be regarded as small spheres arranged in space to form a regular pattern. This pattern can be described in terms of the axes, planes and distances which define various types of crystal system. The cubic system is frequently encountered amongst metals and is shown in Fig. 5.5.

In one type of the cubic system atoms occupy the corners of a cube with one atom in the centre. This particular arrangement is conveniently described as body-centred cubic. Various types of plane can be drawn through such an arrangement, some of which are shown in Fig. 5.5. The diagram makes it clear that the density of atoms varies in different planes, and that the distances between various planes also differ. The directionality of crystal growth referred to in a previous paragraph implies that some of the planes can grow faster than others. When the difference in rates of growth is very marked, as with the intermetallic phases, then the plane type of growth prevails on the atomic, microscopic and macroscopic scales. However, the directionality of crystal growth is not only dependent on the particular crystal symmetries or the bonding forces, which are controlled by the alloy composition, but also on the thermal conditions of crystallisation. In general, the growth planes are those of densest packing; in the case of Fig. 5.5c these are the octahedral planes. Since in most of the metal crystal systems there are two or more planes of close packing, crystals grow by developing several faces. However, other less densely packed planes may grow at the same time. When the heat of crystallisation is removed unidirectionally, the planes that grow fastest are those whose orientations are favourable to the temperature gradient. In the case of body-centred and face-centred cubic metals, for example, the preferred growth direction is that of the cube edge, the $< 001 >$ direction, coinciding with the growth axes of the crystal.

A submicroscopic area of a growing crystal interface is a perfect crystallographic plane in all growth forms. This can develop into a smooth plane of growth on a microscopic or macroscopic scale when all the growth factors are favourable. A more frequent type of growth is the development of multiple planes, described previously (5.4.2) as the cellular and dendritic forms. A full theoretical and detailed explanation of all the causes and variations of the different growth processes is still not available. Instead, normal metallographic tools for revealing the macro- and microstructure of cast metals are used for identifying and classifying the various types of crystal and their typical growth mechanisms. In the present state of knowledge, a systematic and rational list of all growth morphologies and their dependence on alloy composition and on the growth conditions which are encountered in the cast structure of alloys cannot be drawn up.

5.5 ALLOY COMPOSITION, CRYSTALLISATION, RESULTANT
CONSTITUENTS AND CAST STRUCTURE

The fundamental concepts of nucleation and crystal growth apply to all metals and alloys irrespective of their composition. But, as has been explained in previous paragraphs, different metals and alloy phases may not only belong to different crystal systems, but also the morphology of crystal growth may be affected by thermal conditions as well as by the composition of the alloy. Thus, the number of possible types of crystal or grain structure in cast metals is very large, and it is essential to rationalise what otherwise would be a confusing collection of empirical structural data.

As with other properties of cast metals, a good way of summarising the main variations in crystal or cast structures is by reference to equilibrium diagrams. There are about 64 metals in the periodic table of chemical elements, which gives 2,080 binary alloy systems. From the point of view of cast structure, the simplest system is that in which the two metals are completely soluble in both the liquid and the solid state. The appropriate diagram is shown in Fig. 5.6a, and extends over the whole of the composition range.

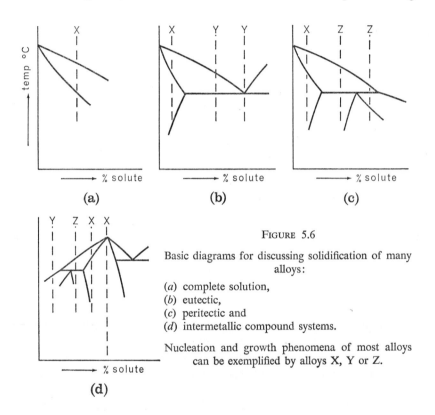

FIGURE 5.6

Basic diagrams for discussing solidification of many alloys:

(a) complete solution,
(b) eutectic,
(c) peritectic and
(d) intermetallic compound systems.

Nucleation and growth phenomena of most alloys can be exemplified by alloys X, Y or Z.

Two other simple diagrams are those of eutectic and peritectic systems, Fig. 5.6b and 5.6c. More complex diagrams include several intermetallic phases of different composition. Such diagrams often consist of combinations of Figs. 5.6a, b and c, as in Fig. 5.6d. It is clear therefore that, from the point of view of crystallisation, a very large majority of the binary diagrams can be discussed in terms of three main types:

(a) One phase only forms during solidification, i.e. pure metals, solid solutions and intermetallic phases. These are indicated by composition X in Fig. 5.6.

(b) Two phases form during solidification. Here several possibilities arise; for instance, one phase may form first, followed by simultaneous crystallisation of two phases as a eutectic, or two phases may crystallise simultaneously from a liquid which is of eutectic composition. There are further possibilities, but all are alike in principle, and are illustrated by composition Y in Fig. 5.6.

(c) A solid phase forms first, reacting with the remaining liquid to give either one or two phases in the cast structure (peritectic reaction). This type is illustrated by Z in Fig. 5.6.

A single binary alloy system may include one or more examples of these three basic examples of crystallisation. The interpretation of the crystallisation mechanism is more involved with ternary or more complex alloy systems, but basically the problems can be understood in terms of the examples outlined above.

Another complication arising in the interpretation of the cast structure of metals at room temperature is that some phases formed during crystallisation may be unstable in the solid state and undergo various types of transformation as the alloy cools. Some of the solid state transformation phenomena that can be used to alter the cast structure to a lesser or greater extent (heat treatment) are discussed in Chapter 6. In the following paragraphs the main types of cast structure which form on solidification and are related to the composition of alloys will be considered. The terms phases, crystals, grains and constituents have been defined in 3.3.2. Some types of impurity which may be present in the cast tructure have been introduced in Chapter 4 and will be further considered in Chapter 6.

5.5.1 Single-phase cast structures

Crystallisation of single phases begins at or below their melting point, i.e. at or below the liquidus line on the equilibrium diagram, depending on the nature of the heterogeneous nuclei present and the thermal conditions, such as the cooling rate and the dynamic condition of the liquid. The number and distribution of nuclei can be further influenced by the growth conditions developed during the solidification, as discussed in Chapter 6. In general, the same conditions of nucleation apply to all the alloys designated X in Fig. 5.6. Similarly, the cast structure of all these alloys at room temperature, if there are no solid state transformations, will only exhibit variations in size and shape of grains (i.e. equiaxed, columnar or a mixture of the two), although the grains will show different orientations of their crystal axes in space, as shown in Fig. 5.7a, Plate 2. The meeting surfaces of crystals, or grain boundaries, which appear as lines in the polished and etched surface, clearly cannot be perfectly matching planes of crystals randomly orientated in space. Furthermore, grain boundaries can be looked upon as yet another example of crystal imperfection in the cast structure as a whole.

In all the alloys considered the crystals may have grown by the plane, cellular or dendritic modes, and the given cast structure, or even a single grain or crystal, may display all the different growth forms at the different stages of growth. Further, the cast grain may not readily reveal the morphology of its growth in metallographic examination.

It is in the compositional features of growth, however, that an important and basic difference arises between the crystallisation of pure metals and intermetallic compounds, on the one hand, and that of solid solution alloys on the other. In the former, the solid deposited on the heterogeneous nucleus and the remaining liquid are of the same composition, and crystallisation progresses at a constant temperature. given slow or equilibrium rates. The total quantities or volumes of the solid and the liquid phases change only as the solidification progresses. With solid solution alloys, on the other hand, the solid deposited on the nuclei has a different composition from that of the parent liquid, as indicated in the equilibrium diagram, Fig. 6.3a. Crystallisation of solid solutions, therefore, involves changes in composition of the solid deposited, as well as of the residual liquid, throughout the freezing process. As such changes are both time and temperature dependent, the growth of solid solution alloys will be much more sensitive to the rate of cooling than that of pure metals or intermetallic compounds. These non-equilibrium effects on cast structure are discussed in Chapter 6.

Although normal metallographic examination at moderate magnifications may indicate that the grains of a single phase may have a substructure due to the micromorphological changes during the growth, electron microscopy and electron and X-ray diffraction methods can reveal other imperfections, ranging from lattice defects down to the larger size submicroscopic defects. For example, the crystal lattice orientation within a single grain may vary by up to 1°; such blocks within a grain are termed mozaics. In solid solution alloys, the solvent grain at the end of solidification is seldom, if ever, completely uniform and homogeneous with respect to the distribution of the alloying elements or solute. This inhomogeneity is termed microsegregation or coring, Fig. 5.7b, Plate 2, and, being a typical non-equilibrium effect, is discussed more fully in Chapter 6.

It follows from this that an overall picture of pure metal, solid solution or intermetallic compound grains in the cast structure is that of an imperfectly grown crystal. The imperfections can be described in terms of irregularities in the atomic arrangement in the crystal planes (lattice defects), inhomogeneous distribution of different solute atoms in the solvent crystal (coring), and microscopic and macroscopic scale imperfections in the arrangement of crystal planes and units relative to one another (substructure). These and other imperfections have an important bearing on the plastic behaviour of crystals in general and of cast structure in particular, as discussed in Chapter 7.

5.5.2 Two-phase and eutectic structures

Although with two-phase structures the same principles of nucleation and growth apply as with one-phase structures, a number of different structural arrangements between the two phases are possible, and a metallographic examination at room temperature may indicate a large number of macro- and microstructural variations. However, all these structures can be classified into two major groups: the eutectics, where the two phases crystallise simultaneously at a constant temperature, and mixed structures containing a single phase with an excess of another phase (or of the eutectic) solidifying over a range of temperatures, Fig. 5.7c, Plate 2. In general, the continuous phase in the microstructure is described as the matrix and the other phase can be described as the dispersed constituent. The terms primary and secondary phases or constituents are used to denote the order of appearance of phases or constituents during solidification.

The problem of nucleation and growth of eutectic constituents of alloys is at the present time one of the least well understood problems of cast structures. As the complete morphology of eutectic structures is somewhat large, the main types will be summarised and then their origin will be briefly discussed. In any single eutectic grain, the continuous or leading phase, or the matrix, can display all the variations in size and shape of crystals analogous in every respect to the crystals of single phases (i.e. equiaxed, columnar, etc.). The dispersed phase, usually discontinuous and smaller in volume, can occur in a number of different shapes as well as in different arrangements of these shapes within the matrix. As a rule lamellar, rod and globular crystals tend to give a regular pattern of distribution within the matrix grain (normal or ideal eutectics), but these as well as more irregular-shaped crystals can gradually degenerate into a more random arrangement in the matrix (abnormal, anomalous, non-ideal or degenerate eutectics), Fig. 5.3b. Metallographic examination of a eutectic alloy can therefore reveal either the size and shape of the matrix grains or the morphology of the distribution of the second phase or both. As a rule, the matrix grains are large (macrosize) in the cast structures and more generally in eutectic studies attention is paid to the various aspects of occurrence of the dispersed phase. Some other structural features within a single grain of a eutectic are discussed in subsequent paragraphs.

Phase rule and thermodynamic considerations as well as simple applications of the equilibrium diagram show that the equilibrium solidification of a eutectic alloy takes place at a constant temperature. With reference to the equilibrium diagram, Fig. 5.6b, it is clear that the nucleation and growth of the solid solutions phase to the left of the eutectic composition proper i.e. if this phase forms first and is likely to be the leading phase) leaves the residual liquid enriched in the solute thus favouring the nucleation and growth of the second phase. When some crystals of the second phase form, favourable conditions for continuing the growth of the matrix phase will again materialise, and in this way a simultaneous growth of both phases proceeds.

Structurally, therefore, the problem of eutectic solidification is that of analysing the factors and conditions of nucleation and growth of both phases and of their interdependence (coupling). The nucleation of the first phase to be deposited from the liquid is governed by the same factors as those discussed for the single phases. The number of heterogeneous nuclei at the required degree of undercooling to initiate their growth and their location with respect to the possible temperature gradient will determine broadly the number and shape of the matrix grains in the eutectic structure. This number can be further altered by the growth conditions as well as by the physical factors (turbulence, vibration or pressures, if applied, 5.6). The nucleation and commencement of growth of matrix grains in the eutectic structure does not therefore raise any problems basically different from those encountered in the solidification of single-phase alloys, and the key problem of eutectic solidification is that of the second or generally discontinuous or dispersed phase and the relation of this to the matrix phase.

The second phase in the eutectic constituent could be nucleated by the primary or matrix phase, or alternatively it may require its own heterogeneous nuclei. In the former case a regular or normal distribution of the second phase relative to the matrix is likely to result (normal eutectic structure), whilst in the second an irregular or anomalous distribution is more probable. However, the nucleation factor alone will not decide the whole structural relationship in a eutectic, since the growth factors of both phases also play a prominent part. Clearly, the liquid in front of the interface of the two growing phases will be subject to the alternating lateral (side) as well as frontal (depth) concentration gradients of the solvent as well as the solute. If the growing eutectic constituent is roughly compared to a multiple sandwich, the composition gradient is clearly carried forward and sideways into the liquid. The growth of the eutectic interface will therefore be governed by all the factors controlling the crystal growth in general, and in particular by the temperature gradients and the growth rates. The ideal or non-ideal arrangements of the two kinds of crystal, as well as the joint crystal morphology of the two crystals (lamellar or others), will also be determined by the joint growth compatibility of the two phases which depends on their relative crystal structures and surface energy characteristics. It is clear from this that, in the growth of a eutectic grain, conditions of growth can be encountered which lead to cellular or dendritic growth patterns within a eutectic grain, Fig. 5.3b. Such cells, if present, being subdivisions within a eutectic grain, should not be confused with the eutectic matrix grain itself.

It is not possible as yet to forecast with certainty, on the basis of atomic or other properties of the two elements forming a binary eutectic, whether and when this will crystallise as ideal and when as non-ideal morphology. Furthermore, a eutectic of two metals A and B can be induced to solidify into different types of ideal or non-ideal growth forms by modifying the conditions of nucleation and growth during the solidification. The data available at present show that in practice one of the most critical factors influencing the type of eutectic crystallisation is the presence of very small amounts of foreign elements or impurities in the alloy. While the mechanism of each particular effect is not clear, this phenomenon is widely utilised in practice to influence the structure of eutectics, as discussed in Chapter 6.

The crystallisation of alloys to the right or left of the eutectic composition does not involve considerations of any features not already discussed. The primary phase forms first as a solid solution or an intermetallic compound, followed by the crystallisation of the eutectic. The nucleation and growth of the single phase takes place as described previously. However, the nucleation and growth of the eutectic may be affected to a varying degree by the excess of a solid phase already in the liquid. Several different structural arrangements can arise in this way, and in general the growth of the first phase in the liquid affects the size, shape and distribution of eutectic grains as well as the detailed structure of the eutectic constituent itself. Of particular interest is when one of the phases of the eutectic grows directly on the existing primary crystals, leaving the other eutectic phase to crystallise on its own. This is a frequently occurring example of a small volume of an ideal eutectic degenerating into a structure which in no way resembles the structure of a eutectic. Other examples are discussed in Chapter 6.

5.5.3 Peritectic reaction

In the two main examples of crystallisation already discussed, the crystals continue to grow from the nucleation stage until the solidification is completed. However, some liquid alloys which begin to crystallise in the usual way until the eutectic temperature is reached are not able to produce the mixed constituent or eutectic proper from the residual liquid alone. The composition of the new solidifying phase β is such that it can only form by reaction of the liquid with some of the solid α crystals already grown. This type of crystallisation is known as the peritectic reaction, and is represented by the formula

$$\alpha + \text{liq} \rightleftarrows \beta$$

where α is the primary phase and β the new crystal phase formed by the reaction. Depending on the original composition, the whole of the α phase may be used up in the reaction and only β crystals remain, Fig. 5.6c. In the alternative case not all of the phase is used up in the reaction and the final structure contains α and β crystals. The mechanism whereby this type of mixed crystal structure is produced is clearly different from that of a eutectic mixture, where both phases grow simultaneously from the liquid. In the absence of such a eutectic or coupled growth, peritectic structure is more like that of a divorced or non-ideal eutectic, Fig. 5.7f. There is some similarity to the eutectic structure, however, in that the nucleation and growth of the two phases are interdependent and, as with the eutectics, the structures of peritectics are very sensitive to the cooling rate. Very often the β phase is nucleated by the already solid α, and in this way the β phase provides an envelope or matrix around the α grains. An example of peritectic reaction is shown in Fig. 3.5, where Fe_δ crystals react with the liquid to give Fe_γ or austenite crystals. Similar peritectic reactions are common in many binary alloys.

5.6 VARIATION OF NUCLEATION AND GROWTH DUE TO DYNAMIC CONDITIONS

Basic concepts of nucleation and growth phenomena, as already described, jointly with the heat flow effects (Chapter 6) are adequate to explain the basic problems of the origin and variations of the cast structure. In addition, however, some dynamic factors may play an important part in the nucleation and growth of crystals. Some of these factors may be incidental to the casting process, as for example kinetic turbulence in the course of normal filling of the mould. In other instances, turbulence may be used intentionally, such as in mechanical stirring or by induced electric current (magnetic field), or when the solidifying liquid is intentionally subjected to vibrational or, sonic impulses or exposed to high pressure.

These various types of application of physical energy may affect nucleation in two ways, by the transfer throughout the liquid of available nuclei through turbulence, or by supplying extra energy for initiating the crystal growth and so activating the heterogeneous nuclei. An example of the former effect is the fracture and transfer of arms of crystal dendrites by mechanical turbulence to initiate nucleation in central regions of the mould. The second effect is produced by high frequency or ultrasonic vibrations, provided these are of sufficiently high energy. Quantitatively, neither of these physical effects has been fully explained, and furthermore physical methods are difficult to apply in general foundry practice for controlling the structure of cast alloys.

Kinetic and vibrational energy may also affect the subsequent growth of crystals. The main mechanism here is the solute homogenisation in the liquid surrounding the growing crystals. The effect is particularly marked with solid solution crystals and with eutectic constituents. In both cases the liquid surrounding the crystals develops a composition gradient in the course of crystallisation, and this may lead to the formation of a barrier impeding further growth

Such a compositional barrier can be dispersed and the liquid homogenised by the application of energy from an outside source. This problem too is closely connected with thermal effects and non-equilibrium conditions of crystallisation and is further discussed in Chapter 6.

Since in most casting processes non-turbulent conditions of crystallisation seldom obtain, it is very important to consider the contribution of this factor to any possible variations and unexpected non-uniformity in the cast structures of a single cross-section, and even more so for a single casting.

5.7 DIRECTIONAL AND ORIENTATED CAST STRUCTURE

Due to the random distribution of nuclei and multidirectional heat flow in typical castings, the orientation of crystals in the cast structure is generally random. As the growth habit of most crystals is preferential in some crystal directions, this property can be combined with that of directional heat extraction during crystallisation, in order to obtain directional, or even any specific, orientations of crystals in cast samples. In the case of single-phase alloys, this implies either single or polycrystalline, unidirectional as-cast structure. With multiphase alloys, the matrix behaves analogously, but the dispersed phase, as for example in eutectics, can be arranged also to grow unidirectionally, thus producing a fibre-like cast structure. As most shaped castings are of too complex design to obtain unidirectional heat flow, the above technique can be applied in practice to relatively simple castings designs only. The related problem of isotropy and directionality of properties of castings is discussed in Chapter 7.

5.8 SUMMARY

The general principles of nucleation and growth of crystals which are pertinent for explaining the origin of the cast structure of alloys are qualitatively well established. Both phenomena are strongly characterised by various kinds of imperfections existing in the liquid or arising in the course of solidification. In nucleation, neither the amount nor the functioning of heterogeneous nuclei can be as yet accurately controlled, but the general principles known enable the direction of change, if not its magnitude, to be affected. The same is broadly true of crystal growth. Thus, in general, the cast structure can be varied within certain limits only, as neither the nucleation nor the growth of crystals can be radically altered once the alloy composition (purity) and the type of mould (cooling conditions) are fixed. The main difference between single and multiphase alloys is that in the latter there is a degree of mutual interplay, in both nucleation and growth, of the two or more phases present. This explains why the cast structure of a multiphase alloy of a given composition can be more influenced by the casting variables than that of a single-phase alloy. In spite of the general limitations in the extent to which the cast structure of alloys can be varied, such changes as are feasible are of great importance in casting metallurgy when considering the problems of improvement and control of properties of cast metals. This is particularly so for alloys whose cast structure cannot be much altered by subsequent heat treatment. The significance of cast structure of heat-treatable alloys, though smaller in comparison with the former example, is still high, since it may control the success of the casting process at the stage of solidification, and because it influences the conditions and outcome of the subsequent heat treatment.

FURTHER READING

TAMMAN, G., *State of Aggregations*, Van Nostrand, New York and London, 1925.
CHALMERS, B., *Physical Metallurgy*, Wiley, New York and London, 1962.
RHINES, F. H., *Phase Diagrams in Metallurgy*, McGraw-Hill, New York and Maidenhead, 1956.
CULLITY, B. D., *Elements of X-ray Diffraction*, Addison-Wesley, Reading (Mass.) and London, 1959.
WINEGARD, W. C., *An Introduction to the Solidification of Metals*, Institute of Metals, 1964.
HURLE, D. T. J., *Mechanism of Growth of Metal Crystals*, Pergamon, Oxford, 1962.
SARATOWKIN, D. D., *Dendritic Crystallisation*, Consultants Bureau, New York, 1957.
TURNBULL, D., The liquid state and the liquid-solid transition, *Trans. A.I.M.E.*, 1961, **221**, 422.
TILLER, W. A., Solute segregation during ingot solidification, *J. Iron and Steel Inst.*, 1959, **192, 338**.
Liquid Metals and Solidification, Symposium of the American Society of Metals, 1958.
SCHEIL, E., Eutectic crystallisation, *Giesserei, Tech-wiss. Beihefte*, 1959 (24), 1313.
CHADWICK, G. A., *Eutectic Alloy Solidification*, Pergamon, Oxford, 1963.
CIBULA, A., Grain refinement of Al-Alloys, *J. Inst. Metals*, 1951–52, **80**, 1.

CHAPTER 6

SOLIDIFICATION: NON-EQUILIBRIUM EFFECTS AND
VOLUME CHANGES

6.1 INTRODUCTION

In the preceding chapter the crystallisation of metallic liquids was considered in terms of nucleation and crystal growth, which were related to those chemical and physical properties of liquids that are directly dependent on the alloy composition. These concepts form the basis for developing a quantitative theory of crystallisation which can then be applied to casting practice when the conditions of solidification are precisely defined. In practice the mould largely determines the conditions of solidification, and the way in which it affects the process of solidification will be considered in this chapter. One of the most important effects of the mould is that it controls the rate of cooling of the alloy during the crystallisation. This, in turn, affects both nucleation and crystal growth. Furthermore, the rate of cooling varies during the progress of crystallisation with consequent changes in the cast structure.

The liquid to solid transformation of an alloy is generally accompanied by a reduction in specific volume, and this can lead to voids and cavities in and amongst the crystals of the cast structure. These volume changes also affect the spatial progress of crystallisation and this frequently contributes to distinct changes in the cast structure, the most important of which are the macrosegregation phenomena. Volume changes may also cause a non-uniform contraction of the solidified casting. Once a continuous solid skin has formed on the casting, the latter begins to contract as a whole, and if this contraction is not uniform over the whole casting, various stress patterns are generated.

6.2 NON-EQUILIBRIUM COOLING RATE: CRYSTALLISATION
OF PRIMARY PHASES

6.2.1 Nucleation

A consideration of the effects of the cooling rate on heterogenous nucleation requires a detailed examination of the origin and formation of heterogeneous nuclei in metallic liquids. Before pouring castings, liquid alloys are heated to 100° to 300°C above their melting points (superheating). Some impurities, originally in the charge or formed in the melting process or introduced by the melt treatments, may still be solid at these temperatures, but others may dissolve completely in the liquid metal. The rate of heating, the length of time for which the higher temperature is maintained, and the rate of cooling to the crystallisation temperature may affect the solution of some possible heterogeneous nuclei among the impurities present. These factors may also affect the size of the nuclei and the way in which they separate out from the liquid alloy solution on cooling. Such diverse changes in impurities, which cannot readily be studied quantitatively, are known as the superheating effect. This implies that the cast structure of an alloy is often sensitive to its history in the liquid state.

The effect of superheating on the cast structure, i.e. size, shape and distribution of all the grains or constituents present, has been determined experimentally for most industrially used metals and alloys. In the great majority of cases the constituents of the cast structure become coarser (all other factors of solidification being constant) if the duration or temperature of superheating is raised. With very few alloys, e.g. some magnesium base alloys, superheating has the opposite effect and the structure becomes finer. In some exceptional cases, like grey iron, superheating may even affect the constitution of the phases formed during crystallisation; the high temperature of superheating promotes crystallisation of Fe_3C, instead of graphite flakes. In these and many similar cases it is not possible at present to explain in detail the exact nature of the changes occurring during superheating. Some experimental evidence suggests, however, that these changes are mainly compositional or constitutional in character and are therefore liable to affect the nucleation, the growth process, or both, during the subsequent crystallisation of the alloy.

The temperature, rate and duration of heating and cooling prior to crystallisation, therefore, are likely to affect the state and number of heterogeneous nuclei in the alloy. The rate of cooling during the crystallisation itself determines the total number of active crystallisation nuclei, i.e. that proportion of the total number of heterogeneous nuclei which will initiate crystal growth. In this way it affects the total number of separate grains in the cast structure. This problem too has so far been studied mainly by direct experiment. The effect of increasing the rate of cooling during crystallisation has been found to change the shape, number and—in some cases—the constitution and structure of the crystals grown. The magnitude of the effect varies mainly with the alloy composition, as shown in Fig. 6.1, and is generally least with pure metals and greatest with the eutectic alloys.

Changes in the rate of cooling can affect the nucleation, the growth mechanisms, or both, and any of these effects may produce similar changes in the cast structure. For example, commercially pure aluminium and many other metals and alloys poured into a metal mould (all other factors being kept constant) give coarse columnar crystals if the pouring temperature is high and very fine equiaxed crystals if the pouring temperature is near the melting point. The relative significance of nucleation and growth phenomena in such an experiment cannot as yet be quantitatively predicted. Application of low pouring temperatures combined with a rapid cooling rate

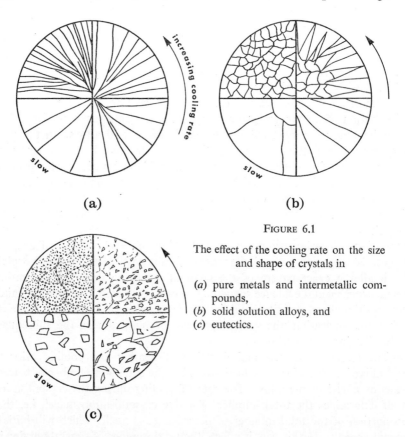

(a)   (b)

FIGURE 6.1

The effect of the cooling rate on the size and shape of crystals in

(a) pure metals and intermetallic compounds,
(b) solid solution alloys, and
(c) eutectics.

(c)

leads generally to reduction in the size of crystals in the cast structure of metals and alloys (6.2.3).

It is interesting to consider the reasons why increasing cooling rate has an effect on the corresponding nucleation rate. Two possibilities arise: either the number of heterogeneous nuclei is further increased, or a larger proportion of the available nuclei behave as active nuclei for crystal growth to occur. Both assumptions are plausible, since a higher cooling rate is accompanied by a lowering of the crystallisation temperature, thus increasing the possibility of more heterogeneous nuclei precipitating out from the liquid alloy. At the same time, some of the nuclei which are inactive at higher crystallisation temperatures may be active at the lower undercooling temperature. The general conclusion from the experiments on the effect of cooling rate on the number of crystals in the cast structure is that the number of potential nuclei in metallic liquids is very much larger than the number of final grains grown, and therefore many potential heterogeneous nuclei remain unused.

Finally, the time and temperature variables could also affect the solute distribution and clustering phenomena in the liquid (3.2.1) and thus influence the nucleation and growth phenomena.

6.2.2 Crystal growth

Nucleation cannot easily be studied independently of crystal growth, but the latter, once it has begun, can be observed in detail experimentally without any further interference from nucleation. This has been done by Chalmers (see further reading to Chapter 5), as illustrated in Fig. 6.2, Plate 3. The solid part of the casting is cooled at one end, a liquid region is obtained by suitable heating in the middle, and a given temperature gradient is maintained at the crystal growing interface. At a given stage of crystallisation, the liquid can be decanted and the growth of the crystal faces examined. With such an experimental technique the effect of the following variables on the crystal growth of alloys can be studied:

C_0 the original concentration of the solute in the melt,

G the temperature gradient at the interface, in °C per unit distance from the interface,

R the rate of growth of the solid (or movement of the interface into the liquid).

One of the most widely used applications of this technique so far has been for studying the effect of a particular type of solute and its concentration on the micro- and macro-crystal growth forms of a given solution alloy, for various conditions of cooling, i.e. variations in G and R values. Several metals with different solute elements that have been examined in this way reveal a general relationship of the type shown in Fig. 6.2b and illustrated in Figs. 6.2c, d and e. The same problem can be analysed in terms of the diffusion laws and heat flow across the liquid-solid interface, giving a graphical relationship between variables of the same type as shown in Fig. 6.2b.

The results shown in Fig. 6.2b were obtained under constant conditions of crystal growth, but such conditions seldom obtain in a mould where neither G nor R are constant during the solidification. In spite of this limitation, several important characteristics of crystal growth emerge from Fig. 6.2b which have a direct bearing on the subsequent discussion of typical cast structures of industrially produced castings.

With a given set of cooling conditions (G/R), an increased concentration of solute C_O leads to a change of crystal growth from predominantly plane to cellular, and finally to dendritic. The data available so far suggest that plane or cellular growth is restricted to low percentage values of C_O and that high temperature gradients combined with low growth rate are required. This finding explains the fact, observed metallographically long ago, that solid solution alloys in practice are largely of dendritic growth morphology. Similarly, for given values of C_O and G, if R is increased, plane and cellular growths are restricted and dendritic growth favoured.

The Chalmers technique of crystal growth analysis has so far mainly been applied to the study of pure metals and solid solution alloys, and is now being extended to eutectics and intermetallic phases. Such work could lead eventually to a quantitative interpretation of the origin of alloy structures in castings. At present it is possible to give only a qualititative interpretation of such structures, not only because of the unknown and variable nature of G and R under the practical conditions of casting, but also due to the uncertainty factor of the impurity effects in industrial alloys.

6.2.3 Combined nucleation and growth effects

It is now proposed to examine the effect of an increasing cooling rate on nucleation and crystal growth under isothermal conditions of solidification, as this is more pertinent to the interpretation of the origin of cast structure in practice. The example to be considered will be that of a solid solution alloy of composition C_O, solidifying in a mould at a non-equilibrium cooling rate, Fig. 6.3a. The general problem of analysing the effect of the cooling rate on other alloy compositions can be treated in an analogous manner.

Crystals begin to grow on heterogeneous nuclei in those regions of the mould which have cooled to the liquidus temperature or below, e.g. at or near the mould walls. The composition of crystals deposited is C_S, so that the composition of the liquid at the growing interface becomes C_O', falling off gradually to C_O in the direction away from the interface, Fig. 6.3b. The existence of this concentration gradient of the solute at the interface has several important effects on the crystal growth, as well as on the final cast structure. Firstly, the magnitude of such concentration gradients is associated with the transition in the growth morphology of solid solution alloys from plane to cellular, and eventually dendritic. Secondly, this concentration gradient is associated with variations in the composition of the solidified crystal. Thus, with a fast cooling rate, diffusion is too slow for crystals of uniform composition to be produced. Such non-uniform crystals have been previously defined as cored, and the phenomenon is described as micro-segregation. The third important effect of the solute concentration gradient is its influence on the rate and progress of crystal growth. The liquid at the interface of the growing crystal has a concentration C_O' and its freezing

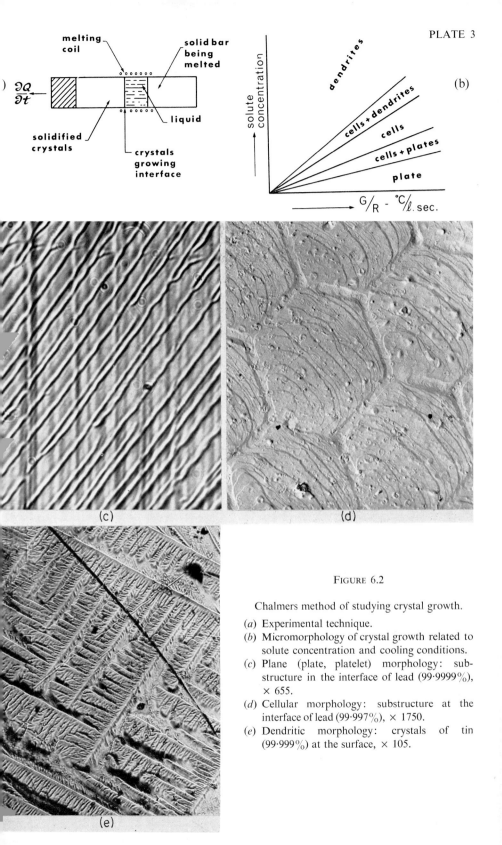

PLATE 3

FIGURE 6.2

Chalmers method of studying crystal growth.

(a) Experimental technique.
(b) Micromorphology of crystal growth related to solute concentration and cooling conditions.
(c) Plane (plate, platelet) morphology: substructure in the interface of lead (99·9999%), × 655.
(d) Cellular morphology: substructure at the interface of lead (99·997%), × 1750.
(e) Dendritic morphology: crystals of tin (99·999%) at the surface, × 105.

temperature is below that of the rest of the liquid, which has a concentration C_0 (see Fig. 6.3c). Consequently three different crystallisation conditions can arise:

(a) The temperature gradient G_1 in front of the interface may be so sharp that the liquid alloy from the region of increased concentration can only grow on to the existing crystals.

(b) The temperature gradient G_2 may be less steep and the temperature of the concentrated region at the interface is below the liquidus

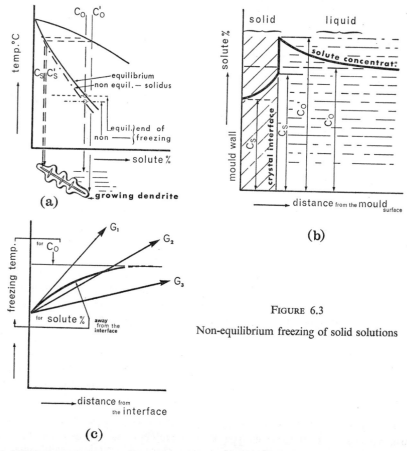

(a)

(b)

(c)

FIGURE 6.3

Non-equilibrium freezing of solid solutions

(a) Solute concentration gradient in the growing crystal.

(b) Solute concentration gradient in the solid, at the interface and in the remaining liquid.

(c) Freezing temperature variation away from the crystal interface as a result of the solute concentration gradient.

temperature of that region, i.e. this region is physically or thermally undercooled.*

(c) The temperature gradient G_3 may be still less steep and both the concentrated region C_0' and the original liquid C_0 may be undercooled.

Conditions (b) and (c) can lead to an independent nucleation and growth of crystals in front of the previously grown interface, the solute-rich region at the actual interface in case (c) acting as a growth barrier against the continuation of growth of crystals forming the interface.

Under condition (a) columnar crystals grow from the mould wall. Similarly, under condition (b) columnar growth will occur if the barrier solute can diffuse away fast enough to prevent the independent nucleation and growth of new crystals in front of the interface. If the barrier persists as under condition (c), then columnar crystals cannot grow, and independent nucleation produces equiaxed crystals throughout the cross-section of the casting. This occurs frequently in practice and a rapid rate of cooling such as is obtained with metal moulds refines the grains of the primary or solution phases in the cast structure, Fig. 6.1b.

6.2.4   Unisothermal cooling

A liquid alloy solidifying in a mould seldom cools at constant G or R values. As the mould temperature is usually much lower than that of the poured alloy, at the moment of contact between the liquid metal and the mould walls, the cooling rate (i.e., R value) may be high, giving an initial chill effect, particularly with metal moulds. When the heat of fusion is liberated, the rate of cooling is slowed down. Finally, towards the end of freezing, when only a small amount of the liquid alloy is left, the rate of.cooling may again increase. Such changes in the rate of cooling or temperature gradients may lead to different types of macro- and micro-structures of solid solution (or other constituents) in different regions of the casting.

One frequently observed type of cast structure in ingots has fine equiaxed crystals at the surface, followed by columnar crystals, which are replaced in the central region of the ingot by coarse equiaxed crystals. Different crystal types are illustrated in Fig. 6.5a. The origin of ingot structure can be explained qualitatively by considering the varying cooling rate and its effect on

---

* Here the physical or thermal definition of the term *undercooling* has been used, but in some textbooks this type of undercooling is defined as constitutional undercooling. This is both theoretically unnecessary and practically confusing, as it implies a different kind of undercooling from the physical. The fact that constitutional changes may occur in front of the growing crystals does in no way justify the application of the same adjective to the undercooling phenomena. The essence of the physical situation discussed is that parts of the liquid in the solidifying alloy may undercool whilst the remainder of the liquid need not.

solidification in the manner described in 6.2.3. It is equally feasible to predict qualitatively conditions of solidification under which an ingot will display one type of crystal only.

Such conditions in practice are still controlled empirically, since the fundamental quantitative data for predicting the nucleation and growth behaviour of cast alloys in moulds are not yet available.

6.3 NON-EQUILIBRIUM COOLING RATE: SECONDARY PHASES AND EUTECTICS

The effects of a non-equilibrium cooling rate on the crystallisation of secondary phases and eutectics can be explained in a similar way to that of solid solution alloys. However, certain additional factors have to be considered.

Secondary phases form in the residual liquid surrounding the primary crystals. The formation of a concentration gradient around the primary crystals, as a result of a non-equilibrium cooling rate, increases the probability of nucleation of the secondary phases in the concentrated regions. Hence these will as a rule be smaller in size than the primary crystals, but their total volume will be larger than that which obtains under equilibrium conditions (6.5.1). As secondary phases are frequently nucleated by the primary phases as well as by heterogeneous nuclei, a rapid rate of cooling leads to refinement of the grain size of the secondary phases. Another important possible effect of a non-equilibrium cooling rate is the replacement of one stable phase by another unstable one. The best known example of this in alloy practice is the crystallisation of Fe_3C instead of graphite in grey irons (6.6).

The effect of the rate of cooling on the cast structure of eutectics is even greater. Nucleation and growth of both phases in a eutectic constituent are so strongly affected by the concentration gradients in the liquid immediately the growth has started that both the micro- and macro-structure of a eutectic can be completely altered. The best known examples are Al–Si and Fe–C eutectics, which are discussed in 6.6.

6.4 EQUILIBRIUM COOLING RATE

By definition, the equilibrium cooling rate of solidification is that characterised by the absence of temperature gradients at the growth interface and in the freezing liquid. The degree of initial undercooling will depend on the nature of the heterogeneous nuclei present, but once the crystal growth has started the equilibrium freezing temperature is re-established throughout the liquid. Crystallisation therefore starts at a number of random points in the liquid, and the uniform temperature results in an absence of concentration gradients of the solute in the solid formed, as well as in the residual liquid. Such conditions favour the growth of the crystals which are first formed and the final structure of a solid solution alloy generally consists of a small number of large equiaxed crystals. The same applies to the secondary constituents and to the eutectic phases. Here too, the equilibrium solidification results in coarsening of the grains, and in some alloys the typical appearance of a eutectic constituent as a regular mixture of crystals may be almost lost. These are described as divorced eutectics.

6.5   NON-EQUILIBRIUM COOLING RATE:

SEGREGATION IN CAST STRUCTURE

The variation in composition of a growing solid solution crystal and of the residual liquid was explained in 6.2, with reference to Fig. 6.3. The consequences of this phenomenon for the homogeneity of solute distribution in individual grains and in a casting as a whole will now be considered.

6.5.1 Microsegregation (coring)

Under non-equilibrium conditions of crystal growth as discussed in 6.2.3, the solute is not uniformly distributed in the crystals. In the example shown in Fig. 6.3 the solute concentration increases from the centre of a crystal towards its periphery. (The opposite occurs with alloys in which a solute increases the melting point of the solvent metal.) This coring or microsegregation of solute was detected by etching techniques in the early days of metallography. However, little quantitative information on the more detailed nature of coring was available until relatively recently, when Castaign developed a technique of electron microprobe quantitative analysis of individual grains. One important fact for which evidence was obtained is that the type of coring varies: it can be gradual, or step-like, or concentrated towards the grain boundaries. Another indication is that the coring, contrary to previous assumptions, in many instances is not eliminated even after a prolonged period of homogenisation in the solid state at an annealing temperature slightly below the solidus (6.9.3). Finally, with reference to Fig. 6.3a it can be seen that, when coring occurs, the freezing range of an alloy is extended and is larger than in the case of equilibrium solidification. This is due to a gradual increase in solute left over in the liquid as the cooling rate increases. In consequence, under non-equilibrium cooling conditions, a secondary phase may form in solid solution alloys which under slow or equilibrium cooling rates would solidify as single-phase alloys. This is an important factor when considering the cast structure of alloys in relation to their equilibrium diagrams (Chapter 3). The same factor is equally important in considering the cause of brittleness, hot shortness (cracking), and other strength properties of the cast structure (6.8). In alloys containing two, three or more kinds of solute atom, each solute will generally have its own coring characteristics. A typical macro- or micro-grain or micro-crystal in the cast structure thus represents a complex unit; not only is it full of structural imperfections in the crystal lattice, but it also has the solute atoms distributed in an irregular manner.

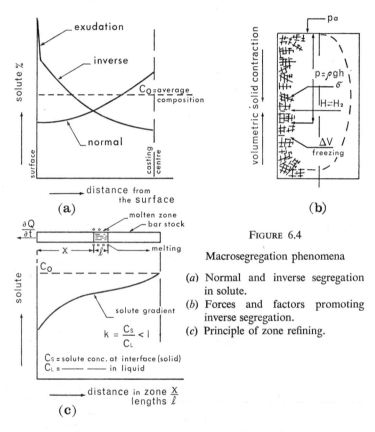

FIGURE 6.4

Macrosegregation phenomena

(a) Normal and inverse segregation in solute.

(b) Forces and factors promoting inverse segregation.

(c) Principle of zone refining.

6.5.2 Macrosegregation (body or major segregation) types

Microsegregation of individual crystals in the basic cause of macrosegregation phenomena within a casting and in the direction of freezing.

As explained above, the solute concentrates normally in front of the growing crystals, and this effect is transported inwards in the direction of solidification. Hence the solute concentration is gradually increased towards the centre of the casting, as shown in Fig. 6.4a. This type of macrosegregation is defined as normal. Clearly the extent of normal segregation is dependent on the extent of coring of individual crystals and hence on the rate of cooling. This kind of segregation frequently leads to an overall solute difference of the order of 0·5% to 1% between the surface and the centre of the casting.

Surprisingly, normal segregation is seldom observed in practice. This is entirely due to the fact that additional factors come into play during the solidification. One such additional factor is the volumetric contraction of liquid metal on freezing. As the crystals grow inwards, so the liquid alloy flows outwards to compensate for the contraction in volume on freezing. This solute-rich liquid becomes even more enriched as it flows outwards, Fig. 6.3a. The average content of solute in the crystals at the outer skin of the casting is thus raised above the value in the original liquid alloy. Therefore, the final macrocomposition gradient in the solute is opposite to that of normal segregation. This is defined as inverse segregation. As the volumetric freezing contraction is of an appreciable magnitude in most industrially used alloys, inverse rather than normal segregation is much more frequently observed in castings. The difference in the amount of solute between the skin and centre of a casting cross-section can be of the order of 1% to 3% of the total solute content. With both inverse and normal segregation, the actual shape of the curves shown in Fig. 6.4a can vary, its form depending on the changes in the solidification sequences of the casting.

The volumetric contraction of an alloy on freezing provides the free space into which the solute-rich liquid flows, resulting in inverse segregation. Several forces may act separately or jointly to change the extent of this inverse flow: the hydrostatic pressure of the liquid metal, the pressure of any previously dissolved gas, the volumetric contraction of the solid outer skin, and the capillary suction force (see Fig. 6.4b). When the conditions are favourable, the solute-rich liquid may flow out through the openings between the surface crystals and solidify at the outer casting surface. Such solidified segregates can be readily recognised with the naked eye on the surfaces of castings, and for different alloys are described as sweat, exudations, blebs or liquation. All these are similar examples of an extreme type of segregation.

In addition to normal and inverse segregations, two other kinds of macro-segregation are encountered in casting alloys, but these are specific to certain alloys rather than of general occurrence. When the crystallising primary phases differ considerably in density from the residual liquid the former may rise to the top or sink to the bottom. This phenomenon, known as gravity segregation, is frequently encountered with some hypereutectic alloys, or with two immiscible liquids (e.g. lead in copper-base alloys). Another type of segregation is important with some cast steel ingots. Here, both gravity and kinetic current forces displace endogenous and exogenous impurities to various positions in the ingot (V and inverse Λ segregates). These kinds of impurity segregation in steel ingots are complex and difficult to explain in detail, as they depend not only on the mechanisms of solidification, but also on the changes in thermal and kinetic conditions occurring in the liquid during solidification.

It follows from the previous paragraphs that in general segregation is more likely to be a severe problem with ingots than with sand castings (6.6.4). Similarly, some alloys are more sensitive to segregation than others. For a simple binary system the tendency to macrosegregation can be expressed in a property diagram relating the segregation tendency to the composition. This follows an almost identical shape to that shown in Fig. 6.10c, i.e. solid solution alloys are prone to segregation.

6.5.3   Zone refining

Although microsegregation in cast structures can be harmful, frequently it is not (Chap. 7); macrosegregation is generally harmful, but can also be put to good metallurgical use. One such application is in zone refining or the purification of metals by means of normal segregation. In a liquid-solid solution alloy which is being cooled unidirectionally, Fig. 6.2a, the chilled end will have less solute than the end last frozen. If some of the solute-rich end of the bar is cut off and the process repeated, the average solute content of the bar is reduced. In zone refining, not all the bar is melted at once, but only a short length of it, and this molten zone *l* is moved progressively across the bar, Fig. 6.4c. By repeating the process, an end portion of the bar of extremely high purity is obtained. The process of zone refining is carried out under controlled conditions, and the main variables of the segregation kinetics can be mathematically related. It can be shown that the solute distribution in zone refining is given by the relation

$$\frac{C_S}{C_0} = 1 - (1 - k)e^{-kX/l}$$

where the symbols are defined as in Fig. 6.4c.

6.6   SOME PRACTICAL ASPECTS OF CAST METAL STRUCTURES

For a given cast product, an important aim may be that of obtaining the optimum cast structure. Some typical variations in cast macro- and micro-structures are discussed below.

6.6.1   Single-phase alloys

In metals of the highest purity, only three types of cast structures are observed, Fig. 6.5a. Single crystals can be grown unidirectionally under a temperature gradient or, less successfully, by equilibrium solidification, on the assumption that since nucleation is a statistical phenomenon there is also a chance of a single crystal growing. By increasing the rate of cooling, finer and finer columnar crystals are obtained, because the resultant undercooling promotes an increasing nucleation. Most crystals in pure metals are so coarse that macroscopic examination is adequate to show the crystal size, and microscopic examination is used to reveal substructural details or imperfections within the grains. It is not as yet possible to obtain very small grain size in pure metal castings by introducing heterogeneous nuclei which do not at the same time appreciably increase the solute content in the metal. Therefore physical methods have to be used to refine the cast structure of very pure metals (5.6).

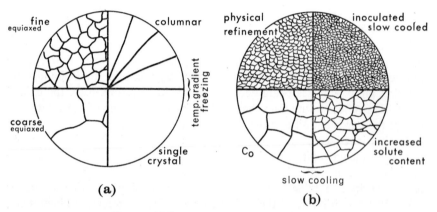

(a)

(b)

FIGURE 6.5 Some variations of cast structure

(a) Pure metals: the coarse structure results from equilibrium freezing, while a fine structure is achieved by physical methods of refinement.

(b) Solid solution alloys: some of the factors affecting relative crystal sizes.

The addition of a solute to a pure metal has two important consequences for the cast structure, Fig. 6.5b. First, with very slow rates of cooling, the coarse macrostructure is gradually refined by addition of solute. This effect is not marked, however, and its magnitude depends on the type and amount of solute added. Secondly, with non-equilibrium and directional cooling, the tendency to columnar growth is gradually reduced as the concentration of solute is increased; when this is combined with a faster cooling rate, columnar crystals are completely replaced by fine equiaxed ones, Fig. 6.5b. Thus by varying the kind or amount of solute and the rate of cooling several types of cast structure can be produced. Two typical casting problems met with in practice are to obtain either a fully columnar or a fine equiaxed structure across a section at very slow cooling rates. For a given type and amount of solute, only a narrow range of cooling rates can induce columnar growth. At present this required cooling rate is best determined experimentally. Obtaining fine equiaxed solid solution crystals in slowly cooled castings has

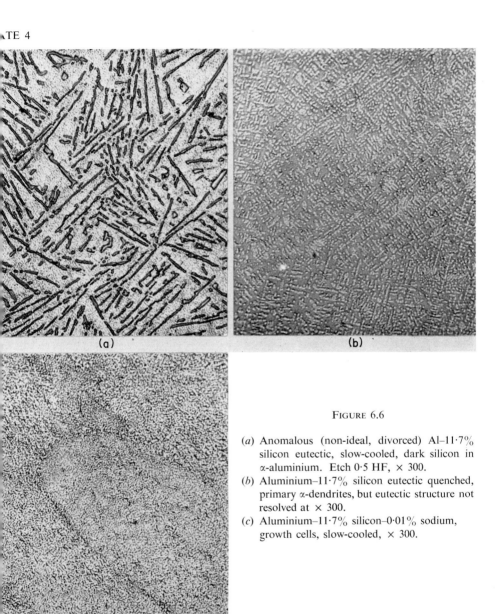

FIGURE 6.6

(a) Anomalous (non-ideal, divorced) Al–11·7% silicon eutectic, slow-cooled, dark silicon in α-aluminium. Etch 0·5 HF, × 300.

(b) Aluminium–11·7% silicon eutectic quenched, primary α-dendrites, but eutectic structure not resolved at × 300.

(c) Aluminium–11·7% silicon–0·01% sodium, growth cells, slow-cooled, × 300.

been a major problem of casting metallurgy. While the solute element itself acts as a moderate grain refiner, this is not sufficient in many cases where the finest possible grains are needed. One answer is to increase the rate of nucleation by adding more heterogeneous nuclei to the alloy (grain refinement by inoculation treatment). For many alloys, however, this problem has not yet been solved on either the theoretical or the practical level. Furthermore, it is not generally possible to separate nucleating effects of inoculants from their possible contribution to growth restriction. For practical purposes, effective inoculating additions have been found for many alloys, as summarised in Table 6.1. Optimum temperatures, time, concentrations and type of inoculant—it may be a chemical compound or a temper alloy, for example—are usually determined experimentally. The same is true of the technique of inoculation (stirring, mixing and timing).

The practical aim of grain refinement is either to improve the strength and plasticity of an alloy during solidification (6.8) or in the finished casting (7.3), or to deal with feeding and soundness problems (Chapter 10). With many alloys such improvements are essential for their industrial applications (e.g., with several magnesium base alloys), while with other alloys the practical advantages of grain refinement are only marginal.

6.6.2 Two-phase alloys: solid solution matrix with dispersed phases

The number of possible variations in the make-up of the cast structure rapidly increases with alloys containing one or more dispersed phase crystals distributed in the matrix. With only a small amount of the dispersed phase present, the matrix or the solid solution phase behaves as it does when this phase solidifies alone, and consequently all the considerations discussed in the previous paragraphs apply. But with each of the different types of matrix structure a number of possible variations in size, shape and distribution of the dispersed phase arise. The most important points concerning a dispersed phase are its size and distribution relative to the matrix—whether, for instance, it has a random or a regular distribution in the matrix grain or at the grain boundaries. Grain refinement by inoculation of the primary matrix phase from macroscopic to microscopic size crystals can also be applied to two-phase alloys. When only the primary phase is inoculated, the grain size of the secondary phase is not directly affected. At present specific inoculation treatments for grain refinement of secondary phases have been developed for very few alloys only. This problem is much more difficult, since the crystals of the dispersed phases are generally microscopic even before treatment. The most successful method used at present is the application of a rapid rate of cooling (illustrated in Fig. 6.1c).

The shape of dispersed crystals presents a special problem. With some alloys only it has been found possible to alter needle-shaped into more equiaxed crystals by small additions of other alloying elements.

TABLE 6.1 *Grain refinement and inoculation additions for structural control of some alloys*

(*Data given are for sand castings. Numerical values of additions may be strongly affected by the cooling rate or other impurities present in the alloy.*)

Alloy system	Phase controlled	Addition of final content % weight (approx.)	Effect on structure	Possible mechanisms
Mg–7–10% Al	α–Mg matrix	Additions: C_2Cl_6 0·05, or $FeCl_3$ 1–2 (the melt may be superheated to 840°C)	Size of grains reduced to 0·03–0·1 mm	Nucleation by Al_4C_3 or $(FeMn)Al_4$
Mg–4–5% Zn	α–Mg matrix	Content: Zr 0·7	Size of grains reduced to 0·03–0·1 mm	Nucleation by Zr or MgZr
Al–4% Cu or Mg or Zn	α–Al matrix	Content: Ti 0·05–0·1 B 0·005–0·01	Size of grains reduced to 0·1–0·5 mm	TiC, AlB_2, TiB_2 nucleation, Ti or B grain growth restriction
Al–5–12% Si	Si eutectic dispersion	Content: Na 0·01	Size of Si crystals reduced to < 0·001 mm	Nucleation and growth of Si reduced
Al–16–20% Si	Si primary dispersion	P 0·005	Size of primary Si reduced to 0·01–0·03 mm	Nucleation by AlP
Cu–35/45% Zn	α–Cu matrix or β crystals	Content: Fe 1–2 ZrB 0·05	Size of or grain reduced to 0·1–0·2 mm	Fe peritectic reaction, ZrB_2 nucleation
Fe–3–4%(C.E.) flake grey irons	Eutectic Fe_3C and eutectic graphite	Additions: FeSi (75%) 0·5 CaSi 0·2 SiMnZr 0·2 SiC 0·2	Eutectic Fe_3C and eutectiferous graphite replaced by normal flakes	Nucleation and growth restriction mechanisms not clear

TABLE 6.1 (*Contd.*)

Fe–4–4·5% (C.E.) nodular grey irons	Flake graphite	Content: Mg 0·04–0·06 Ce 0·005–0·01 followed by one of the inoculants as with grey irons	Flake graphite replaced by spherolitic graphite	As with flake grey irons
Fe–3–3·7% (C.E.) malleable irons	C during malleablising	Additions: B 0·001–0·002 Al 0·002–0·004	Promoting rate of graphitisation in malleablising	As with flake grey irons
	Fe₃C during solidification	Bi 0·01–0·02 Te 0·001–0·002	Preventing mottling in as-cast structure	As with flake grey irons

The distribution of the secondary phase is largely determined by the growth of the primary phase; hence, by refining the grains of the primary phase, the distribution of the secondary phases can be made more even. Physical methods (5.6) affecting crystal growth have also been found useful in improving the distribution of the secondary phase of some alloys.

6.6.3 Eutectic and hypereutectic alloys

The practical problems of the microstructure of eutectic alloys differ to some extent from those of the two alloy groups discussed previously. With a large group of eutectics one of the phases is an intermetallic compound which is as a rule mechanically brittle (Chapter 7). The main problem is therefore that of the microstructure (size, shape and distribution) of the brittle phase, while the matrix grain details are of little practical significance. As many eutectic alloys belong to this group, their industrial usefulness is frequently determined by the degree of refinement and kind of distribution of the brittle phase in the eutectic. If this is continuous or consists of crystals of an unfavourable shape (lamellar, rod- or needle-shaped), the plasticity of the whole structure may be very low, Fig. 5.3b. With a small number of eutectics both phases are ductile and tough, and in such cases the macro- as well as the micro-structure of the whole of the eutectic constituent is relevant for the consideration of the properties of the cast structure.

Typical eutectic problems may be illustrated by considering the three best known practical examples in this field, the Al–Si, (FeSi)—C, and (FeSi)–Fe₃C eutectics.

5

Al–Si eutectic, when slowly cooled, solidifies as an anomalous eutectic, Fig. 6.6a, Plate 4. It has been observed that the silicon phase crystals can be made very much finer by application of a rapid rate of cooling during crystallisation, Fig. 6.6b. Also, even with a slow cooling rate, additions of sodium (0·01%) produce a similar effect, Fig. 6.6c. The grain refinement of the silicon phase is frequently described, without justification, as 'modification,' which implies a change in the nature of crystal growth or a difference in shape of the silicon phase. The available evidence indicates, however, that the silicon grains are merely finer in size and better dispersed by the treatments used; hence the term grain refinement is more appropriate. The fact that either a high rate of cooling or sodium additions produce similar kinds of grain refinement suggests that both nucleation and growth mechanisms are involved. We do not know as yet what part each of these mechanisms plays, because most of the available data on the Al–Si eutectic have been obtained with alloys containing impurities which affect both the nucleation and the growth mechanisms. The number of hypotheses proposed on this topic is large and these are discussed in the literature recommended for further reading.

The author still supports the hypothesis proposed by Ghosh and himself which explains the grain refinement of the silicon phase in the following way. The growth of silicon crystals (almost a pure metal as it dissolves little of aluminium) creates a sharp concentration gradient in the aluminium-rich liquid solution at the interface of the growing crystals. Rapid rates of cooling or the addition of sodium atoms reduce the diffusion rate of silicon atoms through this aluminium-rich concentration barrier, thus promoting the growth of silicon crystals from other nucleation centres. A possible contributory effect of sodium absorption on the growing faces of silicon crystals hindering their growth has not yet been experimentally fully evaluated.

(FeSi)–C (grey iron) and (FeSi)–Fe_3C (white or ledeburitic eutectic iron) can reveal so many different types of macro and micro cast structures that it would require a small book to describe them and a fairly large one to explain them. A brief summary only of the nature of the problem will therefore be given in the present text. Very little has been published on the above problem in terms of the pure iron-silicon-carbon ternary system, as opposed to commercial alloys, since only recently have these elements been available in a high order of purity. Most of the experimental evidence has been obtained by examining the solidification of (FeSi)–C eutectic, containing phosphorous, sulphur, manganese, oxygen, hydrogen and nitrogen in readily measurable quantities, and possibly containing several other elements which are not normally analytically determined. Such alloys show that iron containing from 0·5% to 3% silicon and from 2% to 4% carbon can give a large number of varied eutectic structures, the most representative being shown in Figs. 6.7 and 6.8, Plates 5 and 6.

PLATE 5

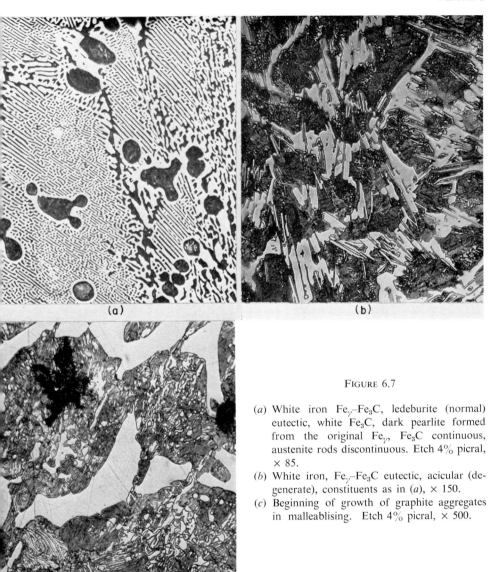

(a)

(b)

(c)

FIGURE 6.7

(a) White iron Fe_{γ}–Fe_3C, ledeburite (normal) eutectic, white Fe_3C, dark pearlite formed from the original Fe_{γ}, Fe_3C continuous, austenite rods discontinuous. Etch 4% picral, × 85.

(b) White iron, Fe_{γ}–Fe_3C eutectic, acicular (degenerate), constituents as in (a), × 150.

(c) Beginning of growth of graphite aggregates in malleablising. Etch 4% picral, × 500.

PLATE 6

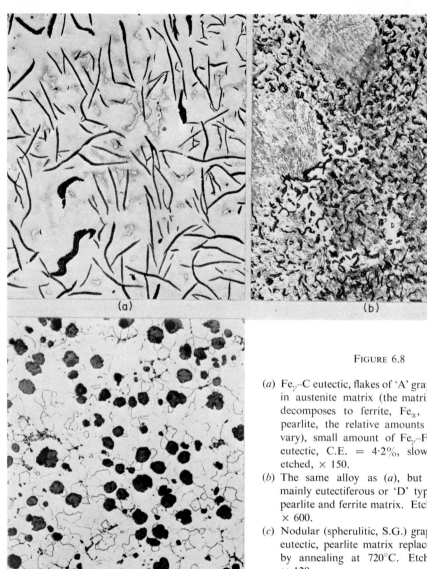

(a)

FIGURE 6.8

(a) Fe$_\gamma$–C eutectic, flakes of 'A' graphite original in austenite matrix (the matrix subsequent decomposes to ferrite, Fe$_\alpha$, and eutecto pearlite, the relative amounts of which c vary), small amount of Fe$_\gamma$–Fe$_3$P phosphi eutectic, C.E. = 4·2%, slow-cooled. U etched, × 150.

(b) The same alloy as (a), but faster coole mainly eutectiferous or 'D' type of graphi pearlite and ferrite matrix. Etched 4% picr: × 600.

(c) Nodular (spherulitic, S.G.) graphite in Fe$_\gamma$ eutectic, pearlite matrix replaced with ferr by annealing at 720°C. Etched 2% nit: × 120.

The two most general features of eutectic structure variations in cast irons are as follows. First, with an increase in carbon and silicon content, the cooling rate being constant, a change occurs with the amount of eutectic Fe_3C or cementite (white or ledeburitic eutectic irons). This at first gradually decreases and then is fully replaced with a flake graphite phase in the eutectic constituent (grey irons). Secondly, at a constant carbon and silicon content, a decrease in the cooling rate also results in a gradual replacement of ledeburitic with graphitic eutectic. Two typical kinds of ledeburitic eutectic, when formed, are the rod and acicular types, Figs. 6.7a and 6.7b, while a fine eutectiferous undercooled or 'D' type graphite and a fully divorced irregular flake or 'A' type graphite are most typical of the grey iron eutectic structures, Figs. 6.8a, 6.8b. Several transitional types of eutectic structure, as well as some entirely different from those shown in Figs. 6.7 and 6.8, are possible in these alloys. Empirical information is available in standard textbooks on cast irons for selecting the alloy composition and cooling conditions required to obtain a casting with any specific type of eutectic structure. It is also clearly possible to obtain a mottled structure, with white and several types of grey eutectic structures occurring in different regions of the same casting. Furthermore, white structures can be transformed into grey structures, Fig. 6.7c in the solid state by an annealing process termed malleablising, according to the reaction

$$Fe_3C \quad \rightarrow \quad Fe_\gamma + C$$

or   iron carbide $\rightarrow$ austenite $+$ graphite

The graphite constituent of the eutectic in normal grey irons can be altered from flakes to nodules or spherulites by desulphurising the melt followed by inoculation (Fig. 6.8c and Table 6.1).

Present knowledge of the metallurgy of grey irons is far from complete. In the solution of the structural problems of the origin of various types of these eutectics and of their transformations, as indeed in many other branches of casting metallurgy, theory has lagged behind empirical methods. However, the latter allow a good measure of day to day controls of most of the practical problems of cast iron metallurgy.

The stable form of free carbon in solid iron-carbon alloys is graphite; but, even with 3% to 4% C in such alloys, the rate of cooling required to obtain the graphite constituent is very slow indeed. Small additions of various elements, for instance silicon, aluminium and nickel (known as graphite promoters), allow graphite crystallisation at more normal cooling rates, while others, such as chromium, titanium, vanadium, manganese and boron (known as carbide stabilisers), retard or inhibit graphite formation.

The major problems in the manufacture of cast irons are:

(*a*) The effect of elements other than iron and carbon on the crystallisation as well as on solid state transformation of iron-carbide eutectic.

(b) The effect of the rate of cooling on the cast structure for a given composition.

(c) The shape variations with growth of the graphite phase, particularly those of flakes and spherulites.

(d) Similar graphite shape variations obtained in the solid phase transformation of cementite by malleablising white irons.

(e) The effects of composition and the cooling rate on the $Fe_\gamma \rightarrow Fe_\alpha$ transformation in the solid state (heat treatment).

The solidification of a eutectic or hypo-eutectic (FeSi)–C alloy can be followed with reference to the Fe–C equilibrium diagram (Fig. 3.5). The first phase formed is Fe_γ or austenite, but the difficulty of simultaneous nucleation of graphite is demonstrated by the fact that the alloy readily undercools. Furthermore, these alloys are very sensitive to the superheating effect, giving white instead of grey irons after superheating to 1450–1550°C. It is likely that only very few of the impurities present would be successful in nucleating the graphite phase, in view of the special type of bonding of carbon atoms in graphite which results in crystal surfaces not readily wetted by other substances. Once the graphite phase is nucleated, since it is 100% pure carbon, a large number of carbon atoms diffusing from the liquids are required for the growth to continue. This is obtained by either a high degree of carbon supersaturation in the liquid, which is brought about by the growth of austenite crystals, or a slow rate of cooling, which allows the carbon to diffuse. The Fe_3C phase, on the other hand, nucleates and grows much more readily partly because of its closer structural similarity with the Fe_γ crystals already present, and also because fewer carbon atoms per crystal volume are required as compared with graphite. The role of inoculation in grey irons is therefore to provide some of the elusive substances capable of initiating the nucleation of the graphite phase. The alloying elements known as graphitisers, such as silicon, may be helpful in nucleation, but it is also evident from studies of malleablising that graphite promoting as well as carbide stabilising elements may have an effect on the growth of the graphite phase. The full physical explanation of all nucleation and growth phenomena is one of the several remaining problems in the crystallisation of cast irons.

It is difficult to explain why the graphite phase grows as flakes in some irons and as spherulites in others. (There is an identical problem in malleablising.) The experimental data available have clearly demonstrated the importance of impurities such as sulphur in this respect. If the molten iron is first desulphurised by magnesium, cerium or calcium, and is then inoculated, graphite nodules grow instead of flakes. As identical inoculants are used for both flake and nodular irons, it is clear that sulphur particularly affects the growth behaviour of the graphite phase. The exact mechanism of this effect is not yet clear. Sulphur also appears to be the critical element for controlling the shape of the graphite crystals in the malleablising process. It is necessary to have some residual iron sulphide in the microstructure of white iron for nodular growth to occur. With an excess of manganese, when iron sulphide is replaced by manganese sulphide, and hence less sulphur is left in the solution, graphite flakes (aggregates) instead of nodules are formed in the malleablising process, Fig. 6.7b, Plate 5.

The complexity of the problems encountered with cast iron structures illustrates the great need for further studies and research in the general field of crystallisation of metals before a full degree of understanding and complete control of cast structure is possible.

6.6.4 Structure and applied aspects of micro- and macrosegregation

Micro- and macrosegregation features in the cast structure affect the plasticity, strength, corrosion and machining behaviour of alloys (Chapter 7). Coring can be detected by X-ray microprobe analysis, etching or heat-treating, or micro-hardness techniques. Segregation of the dispersed phase or of impurities to the preferred crystal plane or to the grain boundaries may result in micro- or macrosegregation forms, and is most readily revealed by etching. Gravity segregation is generally of the macro-segregation type. All types of macrosegregation can be revealed by etching, followed if necessary by quantitative chemical or physical analysis. The sensitivity of an alloy to both micro- and macrosegregation can be indicated in a property-composition diagram (6.5.2), Fig. 6.10c. The extent of micro- or macrosegregation in a binary alloy is greatest in the range of composition of maximum solid solubility, particularly if accompanied by a wide freezing range, and least with compositions solidifying at a constant or narrow temperature interval. The only type of segregation which can be fully or partly corrected by heat treatment is coring (6.9.3). Dispersed phase microsegregation can be removed by heat treatment in alloys which are otherwise single-phase under equilibrium solidification or in alloys which undergo phase transformations in the solid state (6.9.1). Some redistribution of solid impurities and of segregated gaseous elements can also be effected by heat treatment when this allows sufficient time for diffusion to proceed towards the equilibrium conditions. Macrosegregation can only be partially corrected by heat treatment, as it requires impracticably long times, and it is best corrected at the start by controlling the solidification phenomena which cause it.

6.7 SOLIDIFICATION IN A MOULD: VOLUME CHANGES AND FEEDING

6.7.1 Complete definition of the cast structure

In previous paragraphs the cast structure has been considered in simple terms of the crystals or phases or constituents which make up the structure of an alloy. The crystals alone do not, however, define the cast structure fully, and a complete definition includes two distinct groups of defects which generally occur and affect the properties of castings.

In the first group are various imperfections which arise through physical phenomena occurring during solidification (lattice, submicro, micro and macro defects). The origin of these defects has already been discussed, except for submicro, micro and macro cavities. These are due to volumetric contraction or shrinkage during crystallisation. Such cavities are closely related to crystal growth as well as to the pattern of crystallisation of the casting, and are discussed in 6.7.2.

In the second group are various types of process (production) defects, which may originate in the melting, pouring or heat treatment stages of the casting process. These again lead generally to either cavities or inclusion defects in the structure. Purely metallurgical sources of such defects are discussed in Chapters 7, 8, 9 and 10.

6.7.2 Crystallisation front and zones of solidification

When dealing with the problem of volumetric contraction of alloys and its effects on the general cast structure, it is necessary to consider the crystallisation front (growth interface) of all the crystals growing away from a mould wall, and then extend this picture to the casting as a whole. The general case of solidification under a temperature gradient will be considered, and this can then be readily extrapolated to very slow and equilibrium cooling conditions. The three cases of pure metal, solid solution and eutectic alloy represent the most typical examples encountered.

A crystallisation front or interface away from a mould wall is defined as the growth surface of all the crystals growing in the direction perpendicular to the mould wall or parallel to the temperature isothermals. A zone of solidification is the volume enclosed between a cross-section passing through the most advanced points of the crystallisation front and a similar cross-section at the opposite 'all solid' end, beyond which there is no liquid alloy left. The crystallisation fronts and solidification zones of pure metals, solid solution and eutectic alloys are shown in Fig. 6.9. With very pure metals the crystallisation front is generally continuous and is identical in concept with the growth interface of a crystal, the entire growth surface may be flat or slightly corrugated, and throughout the progress of solidification this front is in full contact with the bulk of the residual liquid alloy. Consequently the zone of solidification is either absent or very shallow.

The crystallisation front of solid solution alloys, Fig. 6.3a, is generally different from that of pure metals. Some crystal growth surfaces are in contact with the bulk of the remaining liquid towards the centres of the casting, but in between the crystal dendrite arms, pockets of liquid can be completely cut off from the residual bulk liquid, giving a pasty or mushy zone of solidification,

Fig. 6.9*b*. The depth of the solidification zone in this case is related to the freezing range of the alloy as indicated on the equilibrium diagram and to the rate of cooling. Such zones for a given alloy may be shallow, as is typical in some metal mould castings, but, with slow cooling rates they may extend from the skin of the casting to its centre, as in many sand castings or in large size ingots.

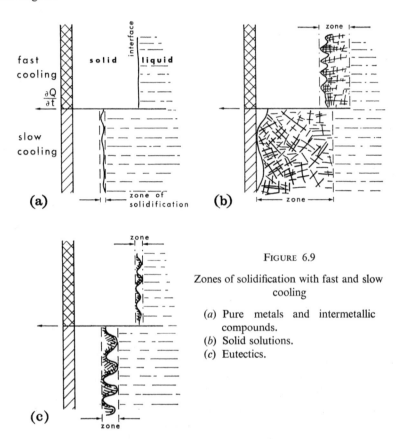

FIGURE 6.9

Zones of solidification with fast and slow cooling

(*a*) Pure metals and intermetallic compounds.
(*b*) Solid solutions.
(*c*) Eutectics.

The crystallisation fronts and zones of solidification of eutectic alloys have characteristics between those of pure metals and solid solution alloys. The crystallisation front is more often corrugated and fan-like in appearance than it is with pure metals. It is generally continuous, but it may contain discontinuous pockets of liquid in between the eutectic grains. Consequently, the solidification zone is generally deeper than with the pure metals and may even be pasty, with areas of liquid enclosed amongst the growing grains or even between the two phases in the grains. This last feature occurs especially with the anomalous or divorced eutectics.

These spatial features of solidifying alloys help to explain how volumetric contraction on freezing could cause cavities in the cast structure. With a continuous crystallisation front, the growing crystals compensate for their volumetric contraction, or feed directly from the adjacent liquid, so long as this is available or the growing front is not cut off from the residual liquid. Isolated liquid pockets occurring during the dendritic growth of solid solution alloys and of fan-growing eutectics in general leave cavities between the crystals when the enclosed liquid is used up. The feeding methods of castings for dealing with these problems are discussed in Chapter 10.

6.8 SOLIDIFICATION IN A MOULD: VOLUME CHANGES AND SOLID CONTRACTION

6.8.1 General

The volumetric contraction of crystals on cooling to and below the solidus temperature creates other problems besides cavities. As soon as a number of adjacent crystals have grown together to take up a distinct shape, i.e. a thin skin at the surface of a casting, then the solid or body contraction of the casting begins, which continues with decreasing temperature. With pure metals solidifying under a slight temperature gradient, this contraction starts as soon as the crystals have joined together and a solid skin with a solidification front has been established. The same is true for solid solution alloys and eutectics, but in this case the body contraction starts while some liquid may still be trapped between the growing crystals.

The mould too can have an important contributory effect on the solid contraction. A casting immediately it solidifies is seldom, if ever, completely free to contract in space, and the mould may hinder a free contraction in several ways. For example, the surface of the casting is not usually completely free to move relatively to the mould surface because of friction or the inter-locking of irregularities of the two surfaces. In addition the casting may also be contracting on to parts of the mould, such as the internal cores. Some of the mould factors controlling the solid contraction behaviour of a casting are illustrated in Fig. 6.10b.

6.8.2 Contraction stresses

The major problem due to solid contraction is that of the formation of contraction stresses. The origin, distribution and types of stress resulting from solid contraction can be explained by reference to Fig. 6.10a. If a casting cools in such a way that there is a temperature gradient from the surface inwards, then at a given time the outer skin has contracted more than a layer deeper below the surface—a situation that extends through the whole of the casting and throughout the period of subsequent cooling to room temperature. Such a non-uniform contraction in a casting as a whole leads, in the example given in Fig. 6.10a, to the inner layer being under compression at the start of the freezing process and the outer layer in tension. As long as the alloy is plastic at high temperatures, these stresses are relieved by plastic deformation (Chapter 7). At lower temperatures, however, the outer parts of the casting start to deform elastically, retaining elastic stresses, a process which later extends throughout the casting. The inner layers still continue to deform plastically at a time when the outer layers have begun to retain the

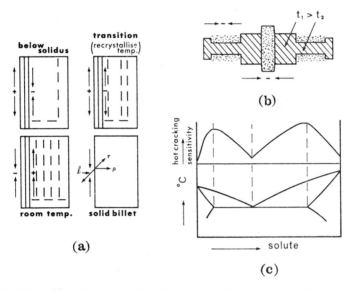

FIGURE 6.10 Solid state contraction phenomena and internal stresses in cast metals

(a) Contraction in a cast billet with falling temperature, resulting in longitudinal (*l*), tangential (*τ*) and radial (*ρ*) stresses.
(b) Origin of stresses resulting from casting design and mould restriction.
(c) Hot cracking tendency (based on empirical tests) related to composition of a binary alloy. Macrosegregation behaves similarly.

stresses. Consequently, when the inner layers begin to retain elastic stresses, as they continue to contract, they will impose compression stresses on the outer layers. This state of stress distribution prevails at room temperatures, and such stresses are defined as residual stresses.

The simple case considered of non-uniform contraction due to thermal gradients in a casting of uniform thickness is but one of many sources of residual stresses in cast metals. Non-uniform cooling due to variations in cross-section of a casting, non-uniform cooling by the mould, as well as various types of mechanical hindrance to free contraction may all produce different types and distribution of stress. The kind of stress (tension, compression, shear) and its direction (longitudinal, radial or tangential, uni- or multi-axial) depends on the casting geometry and the mechanical or contraction factors operating in any given instance. In addition to these macro-stresses or body stresses, there frequently occur in castings residual micro-stresses or tesselated stresses, mainly caused by the non-uniform phase and structural changes in the solid state on a micro or submicro scale. Such stresses are localised, and one example is that of the liberation of gases in the metal in the solid state (4.4.4). On a still finer scale, stress fields around various types of imperfection in the crystals of the cast structure are important when considering plastic deformation and fracture properties of alloys (Chapter 7).

Stresses can lead to three specific phenomena in castings: the appearance of cracks, warping of the casting shape, and the presence of locked-up (internal or residual) stresses which may show up during subsequent use of the casting. Cracking may be on a micro or a macro scale. The term hot cracking is used if it occurs during the solidification at temperatures in the region of the solidus, while cold cracking occurs on further cooling nearer to room temperature. Cracking takes place when the alloy is not capable of plastic deformation and therefore the magnitude of the stress, the composition of the alloy and the mechanism of solidification play important parts in it. Alloys that solidify so that a residual layer of liquid (either the eutectic phase or the impurities) partly envelops the contracting grains are particularly hot short sensitive, as the liquid film leads to cracking instead of plastic deformation. This can be illustrated in a casting property diagram, Fig. 6.10c.

Apart from isolated examples where contraction stresses could be utilised metallurgically, for example in autofrettage (i.e. introducing stresses to oppose the subsequent working stresses), the presence of stresses in cast metals is generally undesirable. The problem of control and measurement of residual stresses is discussed in Chapter 11 but the problem of stress relief is dealt with as part of the general subject of heat treatment.

6.9   HEAT TREATMENT OF CAST ALLOYS

In the present textbook only those metallurgical principles of heat treatment which have particular application to cast metals will be considered. One aspect of heat treatment, concerned with the problems of plasticity and strengthening of the cast structure, is discussed in Chapter 7. Other aspects are briefly summarised in this section.

6.9.1   Alloy constitution, equilibrium diagrams and heat treatment

In the discussion of the solidification process and the origin of cast structure, it was helpful to refer to the liquidus and the solidus lines on the equilibrium diagram. However, most alloy equilibrium diagrams, such as that shown in Fig. 3.5, show that some phase changes can also occur in the solid state. Amongst these the two important ones for discussing heat treatment are changes in solubility with temperature and phase transformations. For example, line DF in Fig. 3.5 indicates that Fe_γ dissolves about 2% carbon at 1145°C and only 0·9% at 723°C. In an alloy containing 2% carbon, the carbon will therefore separate from Fe_γ as the temperature falls to form either graphite at the equilibrium cooling rate or Fe_3C at faster cooling. Similar solubility changes occur in other alloys, Figs. 7.4 and 7.6. Phase transformation changes are almost equally frequent in alloys, and in these a phase stable at a higher temperature changes into a phase or phases stable at the lower temperature. Structurally this implies mainly that one type of crystal lattice changes into another and consequently this may be accompanied by changes in the general grain structure. Such processes occur mainly in two ways: either by the mechanism of nucleation and growth or by a mechanism defined as a block transformation. The former involves a diffusion process, whilst the latter is largely diffusionless. For example, Fe_γ which has a face-centred cubic structure can change into Fe_α, a body-centred structure, by a nucleation and growth process, if the alloy is cooled slowly. The dissolved carbon in this case separates to form either Fe_3C or graphite depending on the carbon content and rate of cooling applied. On the other hand Fe_γ, containing some carbon, can—given a fast cooling rate—transform by a block transformation process into a structure described as martensitic.

Both the change in solubility and the phase transformation phenomena indicated in an equilibrium diagram can be applied in general to alter and possibly to improve the grain structure and hence the properties of various alloys. Such treatment processes are commonly referred to as heat treatments. In general therefore the structure of an alloy formed during the solidification may alter to a larger or smaller extent as the casting cools to room temperature. The term heat treatment thus implies a subsequent specific reheating and cooling process in order to modify one or more of the structural features or conditions of the cast structure which have resulted through the non-equilibrium solidification in the mould in the first place; for example, the removal of residual stresses or non-homogeneity in the structure. Alternatively, heat treatment can be applied to make use of the solubility and phase transformation characteristics of the alloy in order to modify the structure more fully and hence obtain some specific improvements in the structure-sensitive properties. For example, reheating of Fe_α into the Fe_γ field and subsequently back to Fe_α results in a finer structure, i.e. more grains per unit volume, of Fe_α and of pearlite, since each old grain can nucleate and grow more grains of the new phase. This is the basis of the grain refinement of cast steels by an annealing process.

Thus, in general, heat treatment of cast metals can be used to deal with one of the following: (a) removal of some structural weakness resulting through solidification of the original cast structure, (b) redistribution of solute and of the dispersed phases in the matrix to improve some specific properties, and (c) to modify the structure of the matrix or the dispersed phases, or both, to obtain structures having widely different properties from those of the original cast structure. Various examples of heat treatment are considered briefly in this section and others in Chapter 7.

6.9.2 Stress relief

Mechanically, stresses often lead to dimensional distortion of castings (e.g., after machining) or even to cracking, with or without additional external stress. The removal of residual stresses by heating is termed stress

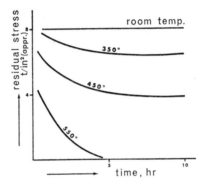

FIGURE 6.11 Typical stress relief time—temperature relation (cast iron example)

relief heat treatment. Structural changes brought about at higher temperatures involve the plastic deformation (flow) of stressed crystals, and readjustment in the uneven distribution of stress fields, with or without measureable changes in the dimensions of the cast component. A stress relief heat treatment cycle involves heating a casting for a given time, at a rate which does not cause additional stresses, to a temperature determined by the constitution and structural condition of the alloy, Fig. 6.11. This is followed by cooling to room temperature at a rate which avoids new residual stresses. In finished castings protective atmospheres may have to be used to avoid oxidation of the metal surface. Although most castings are not stress relieved, with some alloys stress relief is applied as the only heat treatment following the casting process. In other cases stress relief treatment is tied up with other heat treatment processes, for example it may follow a quenching heat treatment process in order to remove the internal stresses produced by it. In an annealing heat treatment, internal stresses in the cast structure are removed at the same time as the annealing temperature is usually much higher than the stress relief temperature.

6.9.3 Annealing

The term annealing, as generally used, covers a number of different types of heat treatment. As applied to cast metals, there are three main annealing processes:

(1) *Structural homogenising anneal* involves heating to below or near the solidus temperature. This modifies structural features produced by non-equilibrium conditions of solidification, such as partial or complete removal of coring or dissolution of secondary phases, or both.

(2) *Solution annealing* or *solution treatment* is used to dissolve a constituent or constituents relatively insoluble at lower temperatures. This treatment therefore usually induces some homogenisation as well. The normal objective of solution treatment is to retain the structure which is stable at the high temperature by quenching to room temperature. Solution heat treatment can be the only heat treatment process applied to a casting, but usually it is followed by ageing or precipitation hardening (7.5.2) or by a stabilising treatment (6.9.5).

(3) *Phase transformation annealing* is applied to those alloy systems, such as Fe–C alloys, where heating to a higher temperature involves a structural transformation of one phase into another. After heating the casting to the transformation temperature, it is cooled to room temperature either slowly (e.g., as in structural refinement by annealing) or by rapid quenching, which may be followed by tempering.

6.9.4 Quenching and tempering

Depending on the rate of cooling and the nature of the alloy, quenching results either in the retention at room temperature of a structure stable at high temperatures or in an intermediate structure which is distinctly different from both high and low temperature stable structures, such as martensite in steels. The martensitic structure obtained by quenching differs significantly from austenite at high temperature and pearlite-ferrite equilibrium structure at room temperature. Tempering implies subsequent reheating following a quenching treatment. The tempering process may lead to hardening (as in precipitation hardening of a solution-quenched structure), but in other instances it results in softening of the quenched structure (as in the tempering of martensite in steels).

6.9.5 Stabilising

In many applications, castings are used continuously or intermittently at temperatures higher than room temperature. Apart from the normal linear expansion, such prolonged heating may lead to permanent dimensional changes, either as a result of stress relief or because structural changes may occur at higher temperatures. In such cases castings can be given a prior stabilising heat treatment, by heating them to and holding them at temperatures high enough to allow these physical changes to take place, and so stabilising the dimensions of the casting. Stabilising may be the only type of heat treatment given to a casting, but more frequently it follows a quenching treatment.

6.9.6 Heat treatment selection

Whether a casting should be heat-treated or not depends on the characteristics of the cast structure of the alloy, the casting design and the operations that follow the casting process.

With some alloys, changes in the cast structure brought about by heat treatment may be essential to give them the properties required. Examples of such changes are the malleablising of white irons, the annealing of carbon steels, the quenching and tempering of alloy steels and solution treatment followed by quenching and ageing of some light alloys. However, some white irons, steels and bronzes in the as-cast conditions have useful properties, for example, hardness and wear resistance. The design and application of a casting may necessitate a heat treatment such as stabilising or stress relief. Subsequent operations such as machining or surface finishing may require heat treatment to improve machining and surface properties. With ingots for subsequent working, heat treatment may be necessary to improve the plasticity of the alloy and hence the deformation properties of the ingot. Thus, in general, a heat treatment is used to introduce such changes in the structure as will either improve the manufacturing operations of the process or will lead to improvement in the properties of castings relevant to their performance in service.

6.10 SUMMARY

In many engineering applications of castings, the required properties depend on their cast structure. This can be influenced to a certain degree by controlling nucleation and growth processes, for example, by means of compositional variations within the specification tolerance (inoculation treatments). The mould in its way affects the cast structure by controlling the cooling rate, which in turn affects the nucleation and growth processes. In many alloys further important structural improvements can be obtained by applying various kinds of heat treatment. Altogether, by controlling the solidification process and by different heat treatments a wide range of cast structures, and hence of properties of some alloys can be obtained. This is one of the more important aspects of casting metallurgy. Present practices of controlling the as-cast structure are based on empirical developments as well as on the application of scientific principles. Heat treatment processes are more directly based on the possibilities of structural change as indicated in the equilibrium diagram. Practical applications of equilibrium diagrams are thus very much more effective for analysing and predicting the changes brought about by heat treatment than for controlling or predicting the original as-cast structure.

FURTHER READING

FLEMINGS, M. C., Solidification of castings, *Modern Castings*, 1964.

KOROLKOV, A. M., *Casting Properties of Metals and Alloys*, Consultants Bureau, New York, 1960.

ADAMS, D. E., Segregation in aluminium copper alloys, *J.Inst. Metals*, 1949, **75**, 809.

The Heterogeneity of Steel Ingots, Reports 1 to 7, Iron and Steel Institute, London, 1929–1937.

PFARR, W. G., *Zone Melting*, Wiley, New York and London, 1950.

Symposium on Internal Stresses in Metals and Alloys, Institute of Metals, London, 1948.

MORROGH, H., The solidification of cast irons, *Br. Foundryman*, 1960, **53**, 221.

MORROGH, H., Progress and problems of cast irons, *Modern Castings*, 1962, **41**, 37.

BURKE, J., Kinetics of malleablising, *Foundry Trade J.*, 1962, **112**, 323.

GHOSH, S. and KONDIC, V. Some aspects of solidification and structure of aluminium-silicon eutectic alloys, *Trans. Amer. Found. Soc.*, 1963, **71**, 17.

SMITH, C. S., Some elementary principles of polycrystalline microstructure, *Metall. Rev.*, 1964, **9**, 1.

GREGORY, E. and SIMONS, E., *The Heat Treatment of Steel*, Pitman, London, 1956.

Heat Treatment of Metals, Institution of Metallurgists, London, 1963.

CHAPTER 7

CAST STRUCTURE AND PROPERTIES OF CASTINGS

7.1 INTRODUCTION

The subject of the properties of castings could be discussed from two different points of view. For the purposes of process control and application of castings, those properties are usually considered which are directly and immediately significant for the functional evaluation of either the casting process or of the cast products. These properties, often described as engineering or practical properties, are discussed in more detail in Chapter 11. On the other hand, many properties of castings can also be examined from the point of view of their relationship to the structure of cast metals. Such properties, described as metallurgical or theoretical properties, are more generally considered in terms of the basic concepts, physical or chemical, directly related to the cast structure. Such a division is largely a matter of arbitrary choice, and in many instances it can be traced to the historical development of metallurgy and of founding. Engineering properties are frequently empirical either because of their complexity or because of their nature. For example, it is difficult to define the property described as hardness of metals, and even more difficult to analyse quantitatively any defined type of hardness in terms of bonding forces of atoms and plastic behaviour of crystals, whilst an empirical scale of hardness measurement and evaluation is both simple and adequate for most engineering problems. On the other hand, the property defined as the elastic modulus is primarily physical concept, the definition, measurement and application of which can serve at one and the same time both practical and theoretical needs. In contrast to the two previous examples, some thermodynamic properties of metal crystals (such as free energy of formation) are more valuable for the theoretical understanding of alloy structures than for dealing with more practical problems.

In general, therefore, the term 'properties of castings' is a comprehensive term and includes both engineering and metallurgical properties, the definition and treatment of which is in some cases identical, while in others either practical or theoretical usage may predominate. Some of the properties are sensitive to variations in either the lattice or the micro- or macrostructure condition of a given alloy as obtained by solidification or by phase redistribution in the solid state; hence the term 'structure-sensitive properties.'

130

Industrially more important amongst these properties are various mechanical, machining, electrical and corrosion properties. Some properties are not structure-sensitive, for example, most thermal properties such as specific heat or linear expansion. In this chapter the more important metallurgical and engineering properties will be examined. Particular attention will be paid to the structure-sensitive properties because of their wider industrial significance. The treatment will be on the following lines. Different structural parameters relevant to property problem considerations will first be summarised; this will be followed by the definition of various properties; and finally the properties will be related to the structural parameters first introduced.

7.2 PARAMETERS DEFINING THE CAST STRUCTURE

Most of the structural concepts which arise in describing the cast structure have been briefly introduced in Chapters 5 and 6. These concepts will now be re-examined from the point of view of their various effects on the structure-sensitive properties of castings.

7.2.1 Atoms

The structure of an atom in terms of its constituent particles determines the forces and energies of bonding of similar or dissimilar atoms in the crystal lattice. For consideration of magnetic and electric properties of alloys it is necessary to examine the electron structure of atoms. The same holds for a theoretical study of the problem of strength of the metallic bond. For metallurgical and engineering purposes, on the other hand, experimentally observed strength and mechanical properties can in general be analysed in terms of lattice or crystal structure parameters. Furthermore, the mechanical properties are invariably determined by direct experiments rather than calculated from the basic atomic properties.

7.2.2 Crystal lattices and their imperfections

Each different crystal or phase in the structure of an alloy has its own characteristic crystal lattice which is determined by the atomic properties of alloying elements making up the phase. Cast crystals as a rule are not ideally perfect and in any individual crystal the following types of imperfections can occur:

(a) Lattice geometry imperfections (point, linear and surface defects).
(b) Lattice distortions due to solute atoms or residual stresses.
(c) Block irregularities due to the presence of crystal blocks (mosaics) of slightly different orientation within one crystal.
(d) Non-homogeneity of ordering or disordering in the distribution of

solute atoms or of discrete sub-microscopic particles (i.e. crystals of another kind) within the main or matrix crystal.

(e) Submicroscopic cavities or voids due either to gases or to solidification crystal growth phenomena (volumetric contraction).

Some of these imperfections are illustrated in Fig. 7.1a, others have been discussed in Chapters 5 and 6. In principle, crystals free from most or all such defects can be grown for theoretical study and even for some industrial requirements (crystal fibres or whiskers), but crystals formed during solidification in industrial processes have most or all of the imperfections described. Metallographic or physical methods are too cumbersome for numerical evaluation of the amounts of such imperfections which are present in crystals. For property measurements, therefore, these imperfections are considered as a characteristic feature of metal crystals, and their effect on the various properties is experimentally determined. The importance of the effect of the imperfections described on the structure-sensitive properties is considered in a subsequent paragraph, but at this stage it is essential to emphasize the specific significance of the linear defects or dislocations. Dislocation imperfections originate in the coarse of crystal growth during solidification and represent different kinds of mismatch which occur in an otherwise regular geometric pattern of atoms in the lattice. Such mismatching can occur as a line of misplaced atoms in crystal planes, when the number of atoms between the two neighbouring planes differs by a factor of one. These are edge dislocations, perpendicular to the plane of the paper in Fig. 7.1a(1). Another type of mismatch is a screw-like displacement in spacing amongst atoms of two neighbouring planes (screw dislocations). In either case the disturbance in regularity of interatomic spacing extends several interatomic distances in the three directions from a dislocation source. The dislocation sources in cast crystals appear to be about 10^2 to 10^3 interatomic distances apart. The number of dislocations (density) and their location are equilibrium characteristics of a grown crystal until they are subsequently disturbed by either thermal or mechanical forces. As a result of the action of such forces, different kinds of dislocation can be interconverted, new sources of dislocation created and old dislocations eliminated. Their main feature is that they provide a mechanism whereby a part of a crystal can glide (slip or shear), relative to another crystal part without the need for bringing into action at the same instant all the intermetallic bonds in the crystal plane along which a particular shearing process is taking place.

When a crystal undergoes a permanent amount of plastic deformation, microscopically this can be observed as the outcome of a process of slipping or gliding of parts of the crystal past one another on a number of crystal planes. However, in reality, and on the atomic scale of observation, the slipping mechanism itself is the net effect of succeeding displacements of consecutive dislocations of the two kinds referred to above. When, for example, a dislocation lying in the plane of a crystal, as shown in Fig. 7.1a(1) extending linearly perpendicular to the plane of the paper, moves transversely (i.e. in the plane of the paper) as a result of the stress, the net result at the outer surface is a small step equal to one interatomic distance in the lattice. A large number of such displacements produce the microscopic effect of crystal slip. As the stress required to start a dislocation moving is very much smaller that that which would produce the same effect in a dislocation-free

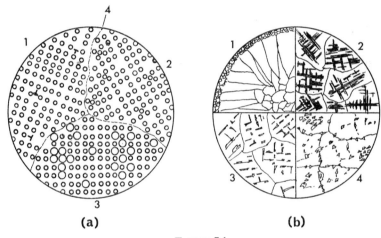

(a) (b)

FIGURE 7.1

(a) Schematic representation of imperfections in crystals: (1) edge dislocations, (2) vacant sites and interstitials, (3) solute atoms: randomness, clustering and ordering, (4) grain boundaries.

(b) Macro and micro characteristics: (1) matrix grain size and directionality (isotropy), (2) solute segregation (coring), (3) preferred distribution of the dispersed phase, (4) randomness and segregation of the dispersed phase.

crystal (the difference of atoms moving one by one rather than a large number at once), the presence of dislocations in most crystals explains why natural crystals deform plastically much more readily than dislocation-free crystals.

Another, but less generally occurring mechanism of plastic deformation of crystals is a block process known as twinning. In this a block portion of a crystal achieves a microscopic displacement relatively to another block of the same crystal when the whole region between the two blocks, i.e. the twin, takes up symmetrical or mirror image orientation relative to the two un-twinned blocks adjacent to the twin. Twinning is more prevalent with some types of lattice structure than others and occurs more frequently under rapid rates of loading and at lower temperatures.

Dislocation slip, and to a smaller extent twinning, are therefore essential factors in the study of crystal plasticity behaviour and of structure-sensitive properties related to plasticity. The effects on plasticity of most of the other lattice imperfections previously described can be understood by the way in which they contribute, or are related to, the dislocation or twinning behaviour, and the same is true of micro-factors (7.2.3). The concept of dislocations, as briefly introduced, will be adequate for a general qualitative consideration of the structure-sensitive behaviour of cast metals, but for a more detailed treatment of the nature, origin and interactions of various kinds of dislocations in crystals, and particularly for a quantitative study of their contribution to plasticity, the texts recommended at the end of the chapter should be consulted. Furthermore, as the present text is concerned with main principles, more attention will be given to the dislocation than to the twinning contribution to plasticity.

7.2.3 Micro- and macro-constituents and defects

Prior to the development of the dislocation theory structure-sensitive properties in general were interpreted in terms of micro and macro characteristics of the structure, i.e., constituent crystals and defects made visible by optical microscopic methods. It is convenient to discuss these structural variables in three main groups:

(a) Number of alloy phases (types of crystal) and the quantity of defects present (non-metallic impurity crystals and cavities).

(b) Size and shape of crystals and of defects (i.e., micro size, $> 0.5 \times 10^{-4}$ mm in optical microscopy, or macro size, >0.1 mm) and equiaxed or elongated shape with either plane, faceted (polyhedral) or irregular boundary surfaces.

(c) Uniformity of distribution of either crystals or defects within the structure, on either micro or macroscale Fig. 7.1b.

In the cast structure of a given alloy these characteristics are a result of the solidification process, which is therefore the first stage of controlling the structure-sensitive properties of castings. Some of these microstructural features can be modified by subsequent heat treatment (6.6), and the resultant degree of improvement in properties determines whether such heat treatment is industrially applied. Many of the properties of cast structure can be explained by considering the inter-effects between the dislocation behaviour and microscopic and macroscopic characteristics of the structure. Before considering such properties some of the more general features of the cast micro- or macrostructures will be summarised, namely, isotropy and homogeneity.

Structure-sensitive properties of any individual crystal are not as a rule isotropic and vary in the different crystal directions. If a casting contains a small number of large crystals only, and particularly if these have a similar orientation, the structure-sensitive properties of the casting as a whole will not be isotropic, Fig. 7.1b(1). Such behaviour can sometimes be favourable and be industrially exploited (directional or fibre structure, 5.7). In the great majority of cases, the number of grains in the cast structure is so large and their orientation is so random that the casting as a whole behaves as an isotropic material. Another kind of anisotropy arises through the uneven distribution of other crystals (dispersed phases and impurities) or of cavities in the main or matrix crystals. If, for example, any one of these is distributed preferentially along certain crystallographic planes or at the grain of the matrix crystals then the properties of such crystals will not be isotropic, Fig. 7.1b(3). Such anisotropy of crystals can affect the properties of the structure of a casting as a whole. Examples of anisotropy will be considered in the discussion of the various properties in subsequent sections.

The term homogeneity of cast structure is used to describe the regularity of distributions of dissolved elements, or of dispersed phases or impurities or cavities, in the structure on either microscopic or macroscopic scale. (The origin of such various forms of non-homogeneity or segregation has been considered in Chapter 6.) Consequently the structure of a casting may not be homogeneous at a given cross-section, and the type or degree of non-homogeneity may vary from one cross-section to another. This is a particularly difficult problem when analysing the properties of a casting as a whole, as the properties at various sections may differ widely depending on the types of segregation present. Examples in this category are discussed in the following paragraphs.

7.3 STRUCTURE-SENSITIVE PROPERTIES, MECHANICAL PROPERTIES

For the reasons outlined in the preceding paragraph it is necessary to consider structure-sensitive properties firstly with regard to that structural condition in which a casting, as a whole, is an isotropic and homogeneous solid. The mechanical properties of castings as defined for engineering purposes can be analysed in terms of the three basic properties of individual crystals making up the structure. These properties are elasticity, plasticity and fracture.

7.3.1 Elasticity

The simplest approach to the concept of elastic behaviour is that of considering the effect of applying to a crystal a uniform tension or compression stress of short duration. Metal crystals tested either as a single or as a polycrystalline body have been shown experimentally to obey Hooke's law in the elastic range, i.e. the stress σ is proportional to the strain ε, the proportionality factor being defined as Young's modulus or elastic modulus E,

$$\sigma = E \cdot \varepsilon$$

In the elastic range of stress, the lattice network of crystals is strained, the internal forces holding the atoms in the lattice opposing the external stress, and on the release of the load the original dimensions of the specimen return. The numerical value of E is a characteristic of a given pure metal crystal the magnitude of which is affected only to a minor degree by solute atoms or by the presence of the dispersed phases in the crystal structure. It cannot be assumed that no disturbance of dislocations has taken place at all during a short-time elastic test, but rather that no permanent measurable microscopic deformation has resulted. However, if the time of the stress application is appreciably longer, and particularly if the elastic tests are carried out at higher temperatures, the strain effects of dislocation movements become permanent and experimentally measurable (creep). Similar but slightly more involved plastic deformation processes may arise if the elastic stress is repeatedly reversed (fatigue), 7.4.5.

7.3.2 Plasticity, ductility and brittleness

After a certain critical stress has been exceeded (defined as the elastic limit of a given alloy), ductile crystals continue to extend, and on the release of stress, while the elastic strain disappears, some permanent or plastic deformation remains. A brittle crystal, on the other hand, fractures either at or soon after the elastic limit has passed. The numerical magnitude of the elastic limit is thus dependent on the onset of an extensive plastic strain operating through the dislocation mechanism. Consequently, several structural features, such as the presence of solute atoms or discrete particles in the matrix crystals, or microscopically dispersed phases, affect the value of the elastic limit of an alloy by increasing the value of the stress necessary to trigger off mass flow of dislocations. The elastic limit is therefore a structure-sensitive property. The difference between ductile and brittle alloy crystals is thus the ability of the former to undergo permanent or plastic strain by the mechanism of transport of dislocations through the crystals. In a brittle alloy, on the other hand, the fracture mechanism is initiated instead. In many cases of plastic deformation it is also necessary to consider the possibility of the twinning process operating.

7.3.3 Strength and fracture

When the stress is increased beyond the elastic limit, several important phenomena occur which have a bearing on crystal strength and fracture behaviour. First, at the localised regions where the individual dislocations have moved, a local increase in stress occurs as the dislocations meet one another either in different planes or at the grain boundaries. The movement of dislocations on their own thus creates a situation in the lattice where a higher external stress is required to produce further dislocation flow. This is a continuous self-hardening process which is described as work or strain or deformation hardening. Further, on the atomic scale, the flow of dislocations is hindered by various obstacles in and around the crystal lattice (sub-grain boundaries, solute atoms and discrete particles and voids), all of which to a varying degree raise the stress required for plastic deformation to continue. The term solution hardening is used to describe the contribution of solute atoms to these strengthening effects. Another contribution to hardening is due to microscopic or macroscopic scale dispersed phases and their boundaries in the matrix (dispersion hardening). Finally, on a macroscopic scale, as a result of the plastic processes described, the change in the shape of the stressed component is also taking place, so that the value of the external stress acting across a unit area of the test piece is also changing.

The appearance of fracture can result from several situations in the stress-strain-crystal lattice conditions, most of which can be explained with reference to a short duration tensile stress. If the load is increased beyond the elastic limit, a structure which is made up of inherently brittle crystals fractures before any measurably plastic strain results. The mechanism of fracture is essentially that of propagation of a fine crack by shearing along crystallographic planes of easiest slip when the shear component of the stress at the root of the crack attains a critical value relative to the stress necessary for plastic deformation. This mechanism implies two sequences: crack formation and crack propagation. Assuming an ideally crack-free crystal, the crack is formed when local conditions in the lattice are reached where the energy required for lattice void formation is less than that for plastic deformation to proceed. In most crystals some cracks are inherently present, due either to crystal growth faults or to phase interfaces and/or grain boundaries, and these can cause some crystals to behave as brittle although without them the crystals would display some plasticity. The difference between brittle and plastic (or ductile) behaviour of crystals is therefore essentially that of the stress and crystal conditions prevailing around the cracks which are inherently present in most crystals. If the stress to move the dislocations beyond and around the crack, or lattice hardening stress σ_h, is smaller than the critical shear crack propagation stress σ_b, the crystal continues to deform plastically. The condition for ductile or brittle behaviour of crystals can therefore be stated in simplified terms of the inequality of the hardening of σ_h and the crack propagation stress σ_b. When the former is smaller, the crystal continues to deform, and vice versa:

$$\sigma_h \lessgtr \sigma_b$$

As the stress to move the dislocations continues to increase due to the various blocking effects and piling up of dislocated regions (work hardening), the stress condition for crack propagation will eventually also be reached in crystals originally defined as ductile. The degree of plasticity of ductile crystals (and conversely the degree of embrittlement) is thus dependent on three major factors: (a) types, amounts and distributions of cracks present in

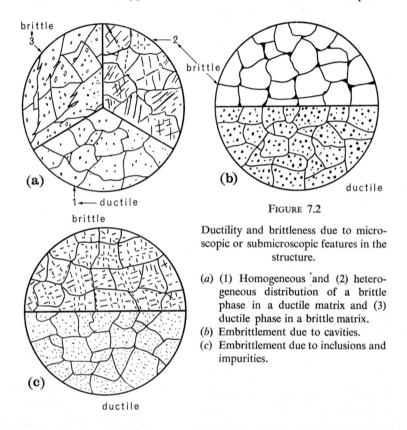

FIGURE 7.2

Ductility and brittleness due to microscopic or submicroscopic features in the structure.

(a) (1) Homogeneous and (2) heterogeneous distribution of a brittle phase in a ductile matrix and (3) ductile phase in a brittle matrix.
(b) Embrittlement due to cavities.
(c) Embrittlement due to inclusions and impurities.

crystals, (b) lattice and micro- or macrostructural features which have a bearing on the movement of dislocations through the crystals and (c) stress required to separate or shear the crystals. For example, the presence of various types of cavities or brittle phases in cast structures can change a ductile behaviour to a brittle one, Fig. 7.2b and c. Similarly, if by controlling the solidification process or by heat treatment a brittle phase, which occurs as coarse dispersed particles in a ductile matrix, is made to appear more finely distributed, such a finer structure provides more free paths for dislocation movement, and the degree of plasticity is improved Fig. 7.2a. Thus, in general, order and disorder (preferred distribution and randomness phenomena in lattice, micro- or macrostructural features have a strong influence on the ductile and brittle behaviour of castings. The dependence of strength-plasticity-fracture phenomena on the factor of time (rate of stress) or the temperature (diffusion processes and stress relaxation) and type of stress (uni- or multiaxial) will be considered in connection with the various types of properties tests.

7.4 MECHANICAL PROPERTIES. DEFINITIONS AND EVALUATION

Functionally, mechanical tests carried out on cast metals fall into three groups: tests for process and quality control (Chapter 11), tests for evaluation of physical constants and property standards for engineering use and application of metals, and tests for analytical evaluation of structure-property behaviour. As in many cases these tests differ essentially in details of apparatus design and in experimental accuracy, only the basic principles of mechanical tests in general will be summarised, with the object of providing a background to the understanding of the mechanical behaviour of the cast structure.

7.4.1 Hardness

Hardness is an empirically defined property which measures the resistance of a material to penetration by another body. A scale of hardness can be obtained by measuring relative sizes of indentation cavities formed by pressing a very hard indentor into the surface of the various materials being tested. The geometry of the indentor, magnitude of load and time of its application are standardised in order to obtain consistent hardness scales. Types of hardness test, their characteristics and uses are considered in the further reading. In an indentation hardness test two basic kinds of deformation occur, elastic and plastic. Structure-hardness behaviour in such cases can be qualitatively interpreted in terms of the concepts already discussed (7.3.1. and 7.3.2). Numerical evaluation or interpretation of hardness data in terms of atomic and lattice structure properties, though feasible for a selected and defined set of measurement conditions, would serve little useful purpose, as the main practical object of hardness testing is that of comparison of materials and process controls. Brittle materials have only a limited plasticity and consequently the selection of indentation conditions is much more restricted, and in some cases an alternative method of hardness testing has to be used (scratch and rebound hardness scales).

7.4.2 Tensile tests

Tensile testing is more widely used than any other method of testing owing to the general importance of strength properties in engineering. In principle, these are simply stress-strain-fracture tests performed on samples subjected to a uniaxial tensile stress. From such experimental measurements some physical as well as some empirically defined tensile properties are obtained. In the former group the main ones are the elastic modulus E, elastic limit and tensile (fracture) strength and, in the other, proof and ultimate tensile strength, per cent elongation and reduction of area. A wide variety of tensile tests is feasible with reference to the size and design of the test piece, rate of stress and temperature of testing. Many of these have been standardised, and present industrial and engineering testing practices are considered in the further reading.

Apart from the numerical value of the modulus of elasticity, which can only be varied to a minor extent by alloying or structural variations within an alloy which do not involve matrix phase transformations, all the other properties measured by the tensile test are structure-sensitive. Proof stress (i.e. stress required to produce a defined amount of permanent strain, e.g. 0.01% to 0.2%) is often used in practice instead of the elastic limit, as it is more readily experimentally measured, Fig. 7.3a. It is an important property for design stress calculations. Ultimate tensile strength, elongation and reduction of area, like hardness, are more useful properties for process control or comparison of materials or for identification of their structural condition than for a theoretical analysis of the strength problem of the metallic state or for design applications. It is for this reason that industrial and theoretical approaches to tensile testing often differ widely. Unlike the hardness test, which is essentially a relative test and furthermore localised to a micro- or macro-region of the surface tested, the tensile test gives stress-strain behaviour representative of the bulk of the material, this being dependent only on the size and location of the test piece used in relation to the whole engineering object made or designed. Structural changes and behaviour of crystals in a tensile test follow broadly the phenomena described in 7.3.1, 7.3.2 and 7.3.3.

7.4.3 Impact

The main object of impact testing is to obtain some practical information on the structural behaviour of alloys when subjected to sudden loads beyond their elastic limit range. Although the values of 'safe' stresses used for engineering design calculations are as a rule smaller than the elastic limit (or proof) stress, it is important to know how a particular alloy would behave if the component were accidentally overloaded. Some ability to absorb overloading by plastic deformation, rather than causing a fracture, is therefore valuable. It can be shown by various mechanical tests that the plasticity or fracture behaviour of crystals is sensitive (apart from those factors already discussed) to the rate and duration of stress and to the temperature level of testing. Furthermore, this sensitivity is also dependent on the particular structural condition of the alloy as defined by its lattice and microparameters. This means that in fact an alloy behaving as ductile in a tensile test may fracture as brittle under an impact load, and such behaviour is dependent on both temperature and structural condition. As the use of ductile alloys predominates in industry, the impact test has been primarily designed to

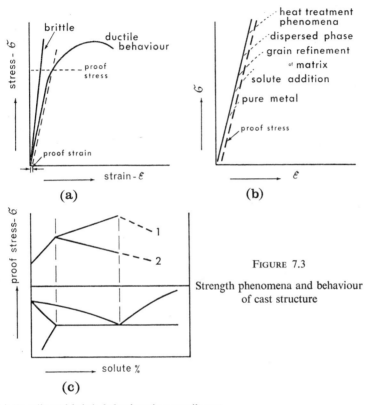

(a) Ductile and brittle behaviour in a tensile test.
(b) Main methods of increasing the proof stress of cast structure.
(c) Strength behaviour of a binary alloy, showing (1) the strengthening effect and
(2) the weakening effect of dispersed phases.

detect and control the structural conditions under which brittle fracture behaviour of an otherwise ductile alloy may be encountered.

The actual impact testing procedure is very simple. The test specimen (notched or unnotched), suitably supported, is subjected to a sudden impact load, usually by a swinging pendulum weight, and the energy absorbed in plastic deformation of the test piece (as measured by the swing of the pendulum beyond the point of impact) is determined. Various impact testing methods used and test design criteria are discussed in the further reading.

In an impact test the specimen is subjected mainly to tensile and compression stresses (at the opposite faces of the actual line of impact). Elastic, plastic and fracture sequences are therefore the same as or similar to that of a tensile test to fracture. What is basically different is that the rate of deformation is very much faster and therefore the sensitivity of an alloy to transport dislocations (or to twin) at a fast rate (plasticity) against the opposing build-up of local stresses and initiation of fracture becomes more apparent. High impact test values of an alloy imply the former and low values the latter condition. This sensitivity is further enhanced by notching the test piece as is often done in standard impact testing procedure.

7.4.4 Creep

In the same way that the impact test has been empirically developed to deal with problems of plasticity and fracture at fast rates of deformation, the creep test aims to deal with the stress–strain–fracture problems encountered in subjecting metals to constant loads over a long period of time. These problems become particularly important when the temperature of application of an alloy is raised and hence, in addition to the purely temperature effect on dislocation mobility of atoms in the lattice, the atoms can also change places by the mechanism of diffusion, which in turn provides opportunity for more dislocation mobility. From the point of view of testing only, the technique of creep testing offers the possibility of several testing procedures:

(a) Choice of simple tension, shear, torque or a combination of these stresses, or even of slowly reversed stresses (of each kind).

(b) Testing at a constant temperature and constant magnitude of stress; by further varying either the amount of stress or the testing temperature, a general pattern of stress—temperature—strain behaviour can be established from which the magnitude of permanent deformation (strain or creep) can be found for any particular combination of stress-temperature and time (creep rate tests).

(c) Continuing the test long enough to obtain fracture (stress-rupture test).

Amongst these different testing possibilities the most widely used in practice is the uniform tension creep test, in order to obtain either the rate of creep or the time to fracture and stress characteristics. The techniques of creep testing, methods of plotting and analysing the creep testing data, and uses of different creep criteria for design and engineering application are discussed in the further reading.

Structural changes and characteristics influencing the creep behaviour of an alloy can be considered with reference to a simple long duration tensile stress. At the application of the load, some elastic extension occurs instantaneously (stage 1). Following this, when the stress is increased, for certain combinations of values of stress and temperature further plastic strain occurs by a dislocation-slip mechanism, but the resultant lattice work-hardening processes eventually cause the cessation of creep (stage 2). However, if stress-temperature conditions are favourable, the temperature factor leads to a partial relief of work-hardening and allows the process of creep to continue. Under these conditions, the rates of softening and hardening are just balanced and the creep proceeds at a constant rate (stage 3). If the applied stress (or temperature) is still higher, localised stress conditions eventually arise leading first to crack formation and then to crack propagation (stage 4). Under a given set of stress-temperature conditions, therefore, stage 1 of creep always occurs, but following this only one or two of the remaining 3 stages are experimentally observed, while the other stages are either not reached or the time of their duration is very short. In a creep test, crystal deformation and hardening processes follow essentially a similar pattern of events and sequences to that in a simple tensile test. The important new factor is, however, that of temperature, which can bring three new phenomena into play:

(*a*) A greater number of crystal planes become available to microscopic slip, which can occur under a smaller stress than at room temperature.

(*b*) Diffusion processes allow dislocations to move more readily from one crystal plane to another, thus providing further facilities for plastic deformation.

(*c*) Localised lattice stress concentration can be relieved, either by diffusion or dislocation strain phenomena.

7.4.5   Fatigue

Fatigue loading differs in two important ways from the examples already considered:

(*a*) A given type of applied stress is reversed relatively quickly and the process repeated (cyclic loading).

(*b*) The maximum value of the applied stress can be higher or lower than the value of the elastic limit of the short duration tensile stress.

In the simplest example of tension-compression stress reversal, the mean stress can be either zero or a tensile or a compressive stress. The other major test variables include the time frequency of stress reversal and the temperature of testing. In any one of the typical fatigue tests, a given type of cyclic stress is applied and the number of stress reversals to failure is obtained. By plotting the peak stress of a cycle (S) against the number of cycles (N), or more usually ($\log N$), a graph indicating the number of 'safe' reversals to avoid failure is obtained. Such a graph can reveal two distinct types of structural behaviour. Some alloys appear to behave as if a definite safe value of stress exists (fatigue limit) below which no fatigue fracture occurs, whilst with other alloys no such clearly defined stress is experimentally obtained. On the whole, many ferrous alloys show such a well defined fatigue limit whilst most of the non-ferrous alloys do not. In engineering practice, therefore, a set number of stress reversals, 10^7 or 10^8, is taken as an experimental target for determining fatigue 'limits' of alloys in general. Available data on fatigue limits of alloys show that the value of fatigue stress which can cause fracture (for the case of a symmetrical tension-compression cycle) varies between 25% to 50% of the value of the ultimate tensile stress. Fatigue testing, similarly to the other tests already discussed, involves the possibility of several different types of test, stress conditions and method of evaluation of fatigue limits. These are discussed in the further reading.

The major structural problem of fatigue is that of explaining the mechanism whereby stresses appreciably lower than that of the short-time tensile fracture stress and even lower than the yield stress can lead to a failure by fatigue. Clearly, in a stress-reversed cycle, some of the lattice elastic, hardening and plastic strain processes occur as with the tensile test previously discussed. However, some new phenomena must also be operative under fatigue, otherwise failure under stresses smaller than the elastic limit stress would not occur. One such phenomenon which has been observed is that the plastically deformed regions under fatigue stress are more selective and are confined to certain localities in crystals only, thus leading to a continuous and gradual pile-up of dislocations in these areas rather than spreading the deformation over the whole of the crystals. Eventually fine cracks are initiated in such regions (first fatigue damage of this kind can be detected in about 10% of the total number of reversals to failure). The fracture occurs in two stages: first it spreads slowly over a part of the test piece, as a rule starting at the surface, and subsequently a stage is reached when the applied stress is capable of spreading the fracture quickly over the rest of the test piece. These two stages of fracture can usually be readily differentiated by visual examination of a fatigue fracture, thus providing ready evidence of whether a component failure was due to fatigue or not. The magnitude of the effects of a large number of variables, such as test piece design or stress and testing condition parameters (notches, surface preparation, presence of initial residual stresses, corrosion action or temperature) on the structural changes occurring under a fatigue load can be measured and largely explained including the structural fatigue behaviour following the first stage of fatigue damage. The main and as yet not fully clarified problem is that of the changes (and their causes) occurring in the structure prior to the first detectable stage of fatigue damage.

7.4.6 Machining

Although a large number of engineering processes involve various kinds of maching operations, there is not as yet a generally accepted machining test which could be used as a quick quality test to assess either the machinability of alloys being machined as affected by their structure or the cutting ability of alloys used for cutting tools. In a simple machining process, e.g. cutting on a lathe, a hard metal tool is held against the component being machined, the force applied on the tool being sufficient to shear off continuously a chip from the component. Even in such a simple cutting process a large number of metallurgical and engineering problems arise; probably the two most important ones are the phenomena involved in the shearing process and the wear of the tool. The former is also related to the surface finish of the component and the latter to the tool life, and furthermore most of the factors encountered in the cutting process are interdependent.

In many cutting processes three distinct stages in chip formation have been observed: plastic deformation in front of the tool edge, shearing off of the chip and, with some alloys, fracturing of the chip into small pieces on leaving the tool. Alloys behaving in the last-named manner are often described as free-cutting alloys. Interpretation of all the structural phenomena occurring in machining is made more difficult by the local rise of temperature which also occurs. It is apparent that the phenomena of plastic deformation, strain hardening and fracture and the effect of temperature on these exert a controlling influence on the cutting process. Some of the characteristics of machining processes are therefore structure-sensitive.

7.5 STRUCTURE-SENSITIVE PROPERTIES IN CAST ALLOY APPLICATIONS

7.5.1 Hardness

By definition, hardness is a localised surface test. Depending on the geometry of the indentor the area affected by indentation may even be within a single grain in the structure. An essential point in the hardness testing of cast structure is, therefore, that of selecting the indentor and the load relative to the particular grain size and constituents, or even defects distribution, of a given cast structure. In other words, a hardness test can be used to measure the average hardness of the structure or the particular hardness of a single constituent, or even parts of it, as grain boundaries (microhardness).

As hardness is a measure of elastic and strain-hardening phenomena of the structure, its application is mainly for testing either the structural condition of a given component or the uniformity of the structure over the component as a whole. The former is based on the fact that yield stress and strain-hardening behaviour are structure-sensitive and therefore a particular structural condition can be identified by its characteristic hardness value. In a sense, hardness provides similar information to the microscopic examination, except that the hardness test is more direct and quick and takes into account a number of lattice and grain imperfections, the effects of which could not be readily assessed microscopically in a quantitative manner. The sensitivity of an appropriately chosen hardness test to homogeneity and isotropy of cast structure is self-explanatory.

7.5.2 Tensile and similar properties

Strength and plastic behaviour of alloys raises two major problems: firstly, variation in properties with the structure of a given alloy and, secondly, comparison of the properties of various alloys. Structural behaviour in the elastic range of all alloys is determined primarily by the nature of the atoms making up the lattice of the matrix crystals. The value of the elastic modulus E can be altered, to a small extent only, by changing either the micro- or macrostructure of a given alloy, or even the alloy constitution, so long as the basic lattice of the matrix crystals is unaltered. Two possibilities of changing appreciably the value of E for a given alloy therefore remain open, either to make use of the phenomenon of crystal anisotropy or alternatively to change the type of matrix crystal lattice by solid state transformations. In the former instance directionally oriented crystals can be grown so that crystal directions having the highest values of E lie in the direction of applied stress. This solution could therefore be applied mainly to castings of simple geometry. The second possibility—that of solid state transformation—could be applied to some alloys only, such as those based on iron. However, in this particular instance the values of E for ferritic or austenitic steels do not differ widely. A completely different approach is that of using the fibre alloying technique. In this case very fine fibres of materials of high E values are introduced into a metal matrix in such a way that the composite behaves as an isotropic solid. This technique is too specialised and expensive for typical casting applications. In general the major emphasis is at present in the direction of control of properties other than the elastic modulus.

The main methods of increasing the proof stress of cast alloys are all based on the principle of introducing such structural obstacles in or around the matrix crystal as raise the stress required for an extensive flow of dislocations in the structure as a whole, Fig. 7.3b. Perhaps the most common of all methods is that of introducing either solute atoms or submicroscopic particles into matrix crystals. Slightly less powerful are the methods of increasing the extent of matrix grain boundaries (grain refinement) or introducing some dispersed phases into the structure. Clearly, such strengthening effects are additive, and can be used either singly or jointly with the others. The greatest strengthening effect can be achieved in alloys in which the matrix crystals undergo phase transformations in the solid state. By rapid cooling or quenching from a high temperature, Fig. 7.4b. various transformation structures, stable at lower temperatures, can be obtained which are highly resistant to dislocation movement. The best known example is the martensitic transformation in some iron and copper base alloys. This strengthening mechanism can also be combined with the previous ones.

From the point of view of proof strength, the structures of cast alloys fall into three main groups:

(*a*) Structures in which the strengthening processes are dependent only on the control of the solidification process (grain boundary and microdispersion hardening by insoluble constituents); many simple solution or eutectic alloys are in this group, Fig. 7.3*c*.

(*b*) Structures in which the condition of soluble phases in the matrix crystals obtained during freezing can be altered by solution treatment and ageing processes in the solid state, Fig. 7.4*a*; many light alloys are in this group.

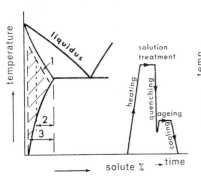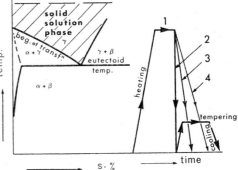

1. Solution treatment temperature	1. Solution treatment (homogenising)
2. Useful range	2. Quenching
3. Composition subject to ageing	3. Air cooling (normalising term for steel
	4. Furnace cooling (annealing term for steel)

(a)   (b)

FIGURE 7.4   Heat treatment strengthening of cast structures

(*a*) Solution treatment, solubility and ageing with typical heat treatment cycle.
(*b*) Solution treatment combined with phase transformation; heat treatment cycle.

(*c*) Structures in which the size of both the matrix grains and the dispersed phases as well as the lattice characteristics can be changed by heat treatment, Fig. 7.4*b*; carbon and alloy steels are the best known alloys in this group.

From the point of view of control of cast structure it is often equally important to reduce the effects of various weakening factors or defects as it is to ensure the positive contribution of strengthening factors. It frequently happens that the former can completely obliterate all the positive effects of the latter.

Amongst the weakening factors in the structure, the three most general ones are cavities, brittle constituents and inclusions. These factors lead to structural weakening by either (*a*) reducing the effective cross-sectional area which carries the load or (*b*) providing localised stress raisers in the structure or even just a simple structural weakness which favours crack formation and propagation. The way in which these factors are interrelated can be seen from the following examples.

Micro- or macro-cavities and brittle constituents with little or no bonds with the matrix crystals (incoherent bonding) cannot transfer stress from one crystal to another, and all the stress is carried by the matrix. The average stress per unit area of the matrix is thus increased by the proportional loss of useful area occupied by cavities and brittle phases. In other words, the magnitude of loss in all structure-sensitive properties is proportional to the defective area present. Clearly, in this case, the direction of the defects in relation to the stress, their shape and distribution in the grain or at the grain boundaries, as well as the continuity of defects are also important. The behaviour of graphite in cast iron provides a good illustration of such effects. Typical flake graphite has a random orientation in the structure, it is incoherent, and discontinuous in the matrix. All the tensile properties of grey iron decrease with an increasing size of graphite flakes. On the other hand, when the size of the flakes is decreased, large numbers of matrix grains are continuous and can transfer the stress from one to another, before the value of crack propagation stress is reached. Much larger improvements in the tensile properties of grey irons are obtained if flake graphite is replaced with spherical or nodular graphite in the structure. The ends of flake graphite in the structure act as small notches, where the local stresses reach much greater values than the average stress owing to the more intense deformation at these locations, thus favouring crack formation.

In general, therefore, cavities, brittle constituents and inclusions can act either as strengthening or weakening factors depending not only on their nature and coherence but also on their shape, size and distribution in the matrix. Their contribution depends almost entirely on the way they affect the dislocation movements and the corollary stress and fracture behaviour for a given type of stress, the rate of its application and temperature. For practical foundry purposes the structural problems of obtaining cavities and non-metallic inclusions to act as strengthening factors are so difficult to solve that a more normal object of cast structure control is to reduce such features to the least harmful conditions. A summary of some typical variations in mechanical properties of alloys as dependent on their structure is given in Table 7.1.

7.5.3 Complex properties
The behaviour of cast structures under impact, creep, fatigue or machining stresses follows some of the general patterns discussed for tensile stresses. This is shown by the fact that the values of these properties can often be related numerically to the tensile test data. On the other hand, the above tests differ from the tensile test in that they bring other factors into play, as already

discussed in 7.4. A common characteristic of all the above tests is that they bring strongly into evidence the phenomena of crack formation and crack propagation and hence these factors become of major importance with respect to the behaviour of cast structures.

In creep testing, structural behaviour at grain boundaries and the thermal stability of both matrix and dispersed phases become important. The higher the temperature, the smaller the strengthening effect and the greater the mobility and hence structural weakening at grain boundaries. Consequently, as a rule, coarse matrix grain size alloys are more creep resistant at higher temperatures. It follows that it is necessary to provide in the matrix one or more of the stable and finely distributed dispersed phases to provide a barrier to the flow of dislocation. Creep resistance temperature behaviour is thus often controlled by the temperature of precipitation and coalescence of the dispersed phases in the matrix.

Fatigue behaviour is also sensitive to grain boundary behaviour. Whilst the strengthening effect of grain boundaries at low temperatures is beneficial, at higher temperatures the opposite holds. For some specific application, as for high-temperature turbine blades, it is advantageous to avoid grain boundaries at the blade surface; hence unidirectionally grown cast structures have an improved fatigue performance at high temperatures. As crack formation plays a vital role in fatigue life, the presence of cavities or brittle phases behaving as cavities at the surface of castings has a detrimental effect.

A particular structural feature revealed by impact testing is the existence of impact brittleness (severe drop in ductility) which is normally not observed by other tests on the same alloy. This occurs either at low temperatures (i.e. below a certain transition temperature) or above room temperature (e.g. blue brittleness of steel). The structural significance of both these phenomena lies in a combination of the speed of deformation with the reduction of mobility of dislocations by dispersed phases or impurities distributed in an unfavourable manner in the matrix.

Specimen preparation, testing and interpretation of data require relatively little time in tensile or impact tests; hence these tests can be used industrially for structural (quality) control or routine tests as well as for evaluation of standard data for design. On the other hand fatigue, creep and machining tests require a much longer testing time and are more generally used for the evaluation of standard engineering data or for the study of plasticity of crystals and of structures.

7.5.4. Segregation and cross-section effects

It has been so far assumed that the general cast structure is uniform across the whole cross-section of a casting, independent of the cross-section size. However, the grain size and shape, the phase and defects distribution in the structure at the surface of a casting may be different from that in the centre.

TABLE 7.1   *Some mechanical properties of metals and alloys as affected by variations in their structure*

Metal or alloy (commercial purity)	Process	Structural condition and variation	Young's modulus $E \times 10^{-6}$ lb/in²	0.1% proof strength ton/in²	Tensile strength ton/in²	Elongation on 2 in. %	Brinell hardness
Al	sand-cast	equiaxed matrix, dispersed impurities	9·9–10	2	5	30	25
	chill-cast	columnar, impurities finer	9·9–10	2	5	40	25
Al— 4% Cu	sand-cast	α-matrix with microdispersion of $CuAl_2$	10–10·5	7–9	10–12	6–8	65–70
	chill-cast	$CuAl_2$ more finely dispersed	10–10·5	8–10	12–13	7–9	65–70
	chill-cast solution-treated	homogeneous α-matrix, no $CuAl_2$	10–10·5	10–12	14–18	8–10	70–80
	chill-cast age-hardened	submicroscopic dispersion of $CuAl_2$	10–10·5	12–14	18–20	4–6	90–100
	wrought-age-hardened	submicroscopic dispersion of $CuAl_2$	10–10·5	13–15	19–22	5–7	90–100
Fe	annealed	homogeneous grains, dispersed impurities	30–30·2	8–9	14–16	35–40	82–100
0·3% carbon steel	as-cast	preferential distribution of α and pearlite	29–30·5	15–16	33–34	10–15	160–180
	annealed	size of crystals reduced, randomness improved	29–30·5	18–20	35–38	15–20	150–170
	normalised	further size reduction of α and better dispersion of pearlite	29–30·5	21–23	38–42	20–25	160–180
	quenched, tempered 500°C	further reduction in size of α and dispersion of Fe_3C	29–30·5	34–36	48–50	15–20	200–220

TABLE 7.1 (*Contd.*)

3·8%–4% C.E. grey irons	normal flake graphite iron	pearlite matrix, dis-persed flakes	18–20	8–11	13–19	0·5	180–200
	nodular, as-cast	pearlite matrix, graphite nodules	22–26	20–26	40–45	2–5	240–290
	nodular, annealed	ferrite matrix, graphite nodules	22–26	13–20	24–32	10–30	130–170
3%–3·7% C.E. malle-able irons	white-heart	ferrite, some pearlite, graphite nodules	27–28	11–13	20–24	4–12	120–220
	black-heart	ferrite, graph-ite as flaky aggregates	24–27	11–13	18–22	6–14	110–150

Similarly thin sections may have a slightly or even greatly different overall structure than thick sections. The same is true of submicroscopic features as well as of cavities and inclusions. Furthermore, within a single grain micro-segregation of both solute elements and dispersed phases can occur and these may also differ in various parts of the casting. All these different segregation features in the cast structure lead to variations in structure-sensitive properties. This is generally referred to as the cross-sectional sensitivity of an alloy. In most instances, cross-sectional non-uniformity in cast structures is detri-mental, but there are certain cases where it is desirable to have a different type of structure at the surface from that in the centre of the casting. For example, hard casting surfaces are required for wear resistance while a tough central core is necessary for load carrying. Such structural features can be produced by controlling the process of solidification, or alternatively by suitable subsequent heat-treatment of the casting (surface heat treatment and surface compositional treatments, for example carburising and nitriding).

The cross-section variations in structure of a casting are dependent on several solidification phenomena (Chapter 6), but in most cases these are simply due to the faster cooling rate of thin sections and the slower and variable cooling rates of thick sections. The best example of cooling rate sensitive alloys are cast irons, as shown in Fig. 7.5a. The slower the cooling rate for a cast grey iron of a given composition, the larger are the graphite flakes, the greater is the amount of ferrite, and the smaller is the amount of

Fe_3C phase in the cast structure. The structure of white irons, on the other hand, must be fully white across a whole section in order to be satisfactory for the subsequent malleablising process (7.6.3). This structure is increasingly more difficult to obtain with an increasing cross-sectional area of the casting, owing to the slower cooling rate favouring graphite formation. Consequently, white irons for malleable iron castings are produced only within a limited range of cross-sections, Fig. 7.5b. In general, alloys with dispersed phases are more prone to cross-sectional variations in structure than solid solution alloys. In some alloys these can be entirely removed by heat treatment, as with cast steels, which show a good uniformity of structure and hence of strength properties across a section, Fig. 7.5c.

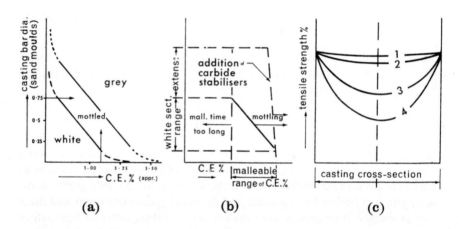

(a)   (b)   (c)

FIGURE 7.5 Cross-sectional sensitivity of cast structure

(a) Cross-section–composition–structure relations with grey irons.
(b) Cross-sectioned sensitivity of malleable irons to casting, annealing and C.E.%.
(c) Examples of variation with cross-section of a mechanical property: tensile strengths of (1) malleable iron, (2) steel, (3) grey iron and (4) tin bronzes.

With cast structures, perhaps the most important example of the cross-sectional structural sensitivity of an alloy is that of the variation in the distribution of cavities across, or with an increasing size of, a cross-section. As the cooling rate decreases with the solidification from the surface of the casting towards its centre, there is a general tendency for both gas and volume contraction porosity to be more concentrated in the central area of the casting, and in the heavier sections. Similarly the slower the overall cooling rate, the larger the volume of the cavities present, since the amount of gas which can be retained in solid solution decreases and the feeding problem becomes more difficult, Chapter 10.

7.5.5 Cast structure, plasticity and ingots

The plasticity of cast structure has so far been considered in relation to the stresses which obtain in typical mechanical tests or during the application of shaped castings. A slightly different problem arises when dealing with the problem of plasticity of ingot cast structures as encountered in processes of mechanical working. Plasticity requirements can be stated in this case in the following terms: the minimum amount of energy should be used to obtain the required semi- or final product shapes, with the minimum number of intermediate steps of deformation, consistent with obtaining the properties desired in the product, as well as the highest process yield. Plasticity of cast ingot structure implies therefore the ability of shape forming without internal or external cracking on a submicro, micro or macro scale. In many instances some edge and/or even surface cracking is allowed for on economic grounds, as such process defects can be removed whilst still retaining the main product undamaged.

Most of the plasticity concepts already introduced—structural features, temperature, rate, magnitudes and kind of stress—play a part in the plastic working of ingots. The major new factor is that of frequent non-uniformity of stresses or temperatures as dictated by the kind of working process used. Thickness depth and edge effects in forging, edge effects in rolling and surface effects in wire drawing are typical examples. The major problem is therefore that of ensuring an adequate plasticity of the initial ingot cast structure to make a particular combination of stress-temperature-rate of the subsequent deformation process technically feasible. A number of examples can be given to illustrate typical problems encountered in ingot working.

The extent of plasticity of an alloy is primarily controlled by strain-hardening phenomena (7.3.3). Consequently most ingot working processes are carried out at sufficiently high temperatures to reduce the extent of strain hardening to a minimum (hot working or working above the recrystallisation temperature). Cold working is more generally used for highly plastic alloys or for the work finishing processes of alloys to obtain the required properties. Some of the embrittling features in the cast structure (cavities, brittle phases and inclusions) may necessitate careful working at the start (breaking down of the cast structure) and then, with a gradual improvement in plasticity, the working process is more freely utilised. Most, if not all, structural weaknesses of the original ingot cast structure can be removed in this way (welding of cavities and breaking down of brittle phases into smaller particles). A combination of working and reheating has often to be used to control work-hardening and recrystallisation phenomena. Tolerance of the extent of breakdown or of casting defects in the cast structure is greater with alloys which do not work-harden rapidly. Rimming steel, which in ingot bulk has an almost spongelike structure but a sound skin, can be hot-worked readily, whilst alloy steels or strong light alloys require as perfect a cast ingot structure

TABLE 7.2 *Summary of ingot structure and metal working variables*

Cast structure, plasticity factors	Deformation conditions	Property requirements	
		Ingot structure	Semi- or finished products
Matrix	Temperature (hot or cold working)	Structural plasticity	Surface quality
Dispersed phases	Stresses	Isotropy	Mechanical properties
Cavities and inclusions	Rate of deformation	Homogeneity and segregation	Isotropy

as it is technically feasible to obtain. Anisotropy of crystals in ingots (e.g., long columnar crystals) as well as anisotropy in orientation of the worked structure (preferred orientation) are sometimes special problems in ingot working metallurgy. A general summary of the major variables of ingot working is given in Table 7.2, where each variable in the first column is related to all the variables in the remaining columns.

7.5.6 Other properties of castings

In addition to the mechanical property requirements of castings a number of other properties may often play a significant and even decisive part in the selection of a casting alloy for a given application. These properties will only be briefly summarised, as their full discussion requires considerations of principles not otherwise discussed in the present book.

(*a*) *Corrosion resistance.* The controlling factor in the liquid corrosion resistance of an alloy is the electrochemical character of the phases in the structure, which determines the corrosion behaviour of the alloy in a given corrosive medium. Variations in the phases of cast structures may influence the degree of corrosion resistance: for example, a constituent formed at the non-equilibrium cooling rate may lower the corrosion resistance, while the corrosion behaviour may be improved by an annealing process. Both soluble and insoluble impurities and cavities may considerably affect the corrosion behaviour of an alloy. Corrosion resistance is also strongly influenced by surface films produced naturally, artificially or by the corrosion process itself (e.g. oxides). Stress corrosion is particularly important as it affects crack formation and propagation.

(*b*) *Joining processes.* Castings are frequently welded, brazed or soldered for various applications. These processes are also often used for the repair of castings or in the assembly of simple castings to produce complex shapes which would otherwise be difficult to cast. In general, in joining problems the basic composition (and hence the corresponding structure) is the main factor, but cast structure variations can also be important. For example, joining processes often involve localised heating of the parts to be joined, resulting in stresses. Plasticity, which is a structure-sensitive property, as discussed in 7.3., can therefore be important. Structural differences between the weld and the casting are also of major significance.

(c) *Finishing processes.* Cast surface finish is an important feature for some castings, particularly when some finishing process such as polishing, electrochemical and other functional or protection treatments is subsequently applied. In general, cast surface smoothness is governed by moulding and casting variables, (Chapter 8), as well as by basic alloy composition. However, the cast structure may also be significant. Crystal coarseness, or the presence of hard and brittle dispersed phases or cavities and voids at the surface due to segregation and impurities, are examples of this.

(d) *Damping capacity.* Some solids vibrate when subjected to a sudden impact stress below the elastic limit. The ability of a given structure to damp such vibrations by converting them largely into heat energy is defined as the damping capacity. This is a structure-sensitive property, the damping capacity of flake grey iron being very high and that of nodular iron much lower. In other words, nodular iron gives a loud, bell-like sound when hit, but grey iron does not. The mechanism whereby a given structure achieves the dampening effect varies: diffusion, stress relaxation, thermo-elastic and magnetic phenomena as well as dislocation movement.

7.6   STRUCTURE AND STRENGTH PROPERTIES OF SOME INDUSTRIAL ALLOYS

The nomenclature and industrial classification of cast alloys is, at present, not based on scientific concepts. The main reason for this is that useful alloys were developed mainly empirically and over a long period of time. The final choice of alloy composition in industry is often a compromise between conflicting factors, such as properties of the casting required (Chapter 11) casting properties of the alloy (Chapter 6) and the economic factors of materials and production. Various kinds of classification of alloys are summarised in publications of standard specifications (Chapter 11) which are compiled for industrial and functional rather than for scientific needs. It would be costly, though in the long run more economic, to change the methods of alloy classification at present used in industry. At best these methods are confusing and vary from trade to trade, or from one authority to another, and from one country to another. By contrast, the scientific problem of systematic classification of industrial alloys would be a relatively simple one. Excluding production cost and structural defects, the main criterion on which the properties of cast alloys are dependent is their structure. In general, this consists of the matrix structure of a particular element in which other elements are either dissolved or form dispersed phases. A consistent and comprehensive system of alloy classification could thus be built on a structural basis. The historical development and current usage of some typical industrial alloys are reviewed in the subsequent paragraphs.

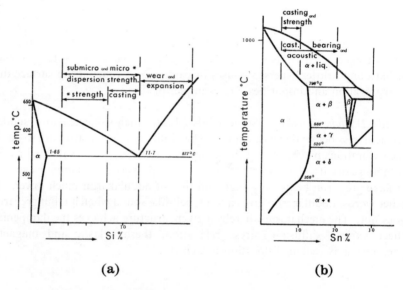

(a)   (b)

FIGURE 7.6 Properties and cast structure of industrial alloys

(a) Al–Si alloys ; additional alloying elements may be used to raise the level or alter properties; the ductile and brittle phases can be varied by composition, casting and heat treatment.
(b) Cu–Sn alloys (bronzes).

7.6.1 Aluminium–silicon alloys

An examination of industrial alloys based primarily on this binary alloy system reveals that the alloys fall into three groups: hypo-eutectic, eutectic and hyper-eutectic, Fig. 7.6a. Simple binary alloys are based on strengthening aluminium crystals with solute silicon atoms and by microdispersion of the second phase, essentially made of hard but brittle silicon crystals. The most widely used are the eutectic alloys, on account of their good casting properties (7.7) and good combination of strength and other functional properties. The strength behaviour of a eutectic alloy is largely controlled by the fine particle size dispersion of the brittle silicon phase in a ductile α-aluminium matrix. Hence the need for grain refinement of the silicon phase by either sodium addition or a rapid cooling rate, or both. Hypo-eutectic alloys contain fewer silicon crystals, but the relatively soft α-aluminium matrix can be age- or precipitation-hardened if the alloy contains a small amount of magnesium. The intermetallic Mg_2Si crystals can be dissolved and dispersed in the matrix by the mechanism shown in Fig. 7.4a, thus forming an alloy stronger than any one of the eutectic binary alloys, but which still has good casting properties. This is an example of adding alloying elements to simple binary alloys in order to improve some of their specific properties. Small magnesium additions can be made to the eutectic alloy to get a heat-treatable alloy similar to the hypo-eutectic alloy. On the other hand, small additions of nickel and copper give low expansion characteristics on heating, important when used for pistons. Magnesium can also be added to make the alloy heat-treatable. Hyper-eutectic alloys are structurally similar to the eutectic alloy except that the matrix contains the dispersed eutectic silicon as well as the primary silicon crystals. Similar properties to the eutectic alloy, but with an increased wear resistance, can thus be obtained.

The strength properties of most light alloys in general are based on the principle of solution strengthening, frequently augmented with submicroscopic dispersion strengthening (age-hardening) for high-strength alloys and microscopic dispersion in the as-cast structure for moderately strong alloys. The application of the alloy determines the choice of the strengthening mechanism and the additional alloying elements, so that strength is combined with one or more of the other properties required for the manufacturing process or for the functional use of the casting. In many cases such a combination of properties can be achieved only by using several alloying elements.

7.6.2 Copper–tin alloys, or bronzes

Bronzes were among the first alloys used by man. Yet, in spite of their long history, the basic composition of bronzes has changed very little. Crystals of copper form the matrix of the cast structure, and these are strengthened by a solution of tin in copper. At the limit of solid solubility, Fig. 7.6b, another phase β forms, by the peritectic reaction, but this phase is only stable at higher temperatures and changes at equilibrium, first into γ and σ phases and later into the ε phase. With the normal cooling rates, however, the σ phase is stable at room temperature. The causes of the increased instability of some alloy phases with a decreasing temperature can be explained in terms of structure-energy stability data in alloys. From the point of view of properties or microscopic structural analyses of bronzes, however, the significant feature emerging from such phase transformations is that the original β crystals of a certain characteristic lattice structure are replaced by σ crystals of different structure, but the size and distribution of these crystals in the matrix at high temperatures are similar to those at room temperature. When considering temperature dependent properties, such phase transformations have to be taken into account.

Structurally, therefore, tin bronzes at room temperature resemble aluminium–silicon alloys. A brittle intermetallic compound, the σ phase, is microscopically dispersed in the ductile matrix of the α-copper crystals, but the strengthening behaviour of the two alloys is different. Unlike aluminium–silicon alloys, tin bronzes develop their maximum strength by solution hardening of tin in copper, and the appearance of σ leads to a gradual fall-off in strength. This is largely due to the metallurgical difficulties of controlling the microscopic dispersion of σ in bronzes, where, apart from a rapid-cooling rate, no other metallurgical means of refinement and dispersion of the σ phase has been found in practice.

Consequently bronzes with a high proportion of the σ phase have a very limited application. Also no tin bronze alloys yet used widely utilise the principle of the submicroscopic dispersion hardening in the α copper phase, although such a strengthening mechanism could be applied. The reason for this is that the most important applications of bronzes depend on corrosion resistance or on frictional properties (bearings), or acoustic and colour appeal, rather than on strength behaviour. The optimum types of cast structure for such varied applications are less stringent than those where strength is a major consideration.

Another contrast between the tin bronzes and aluminium–silicon alloys, from the casting point of view, is that the former have a long and the latter a short, freezing range. With reference to Fig. 3.7c, it can be readily seen that bronzes are difficult casting alloys on account of their sensitive to segregation and cavity formation.

Consequently the ternary and more complex types of bronze have been developed, by addition of zinc, lead and nickel, to counteract and deal with such casting problems rather than to control the strength behaviour.

7.6.3   (FeSi)–C alloys.   Malleable and grey irons

Structurally, steels, malleable and grey irons resemble the previous two alloy groups in being made up of a mixture of ductile matrix and brittle dispersed crystals. A variety of methods for strengthening the matrix, combined with a great number of possible arrangements and conditions of the brittle phases, make these alloys structurally the most varied and, as regards their properties, industrially the most useful of all alloys. Their Young's modulus, high melting point, structural response to various alloying additions to obtain specific properties, and the relative abundance of iron ore in the world, combined with the ease of separation of iron from its oxide and the extensive empirical knowledge of its refining and melting, explain why a whole period in history is described as the iron age.

The most widely used are iron–carbon and iron–silicon–carbon alloys and these can be directly related to the equilibrium diagram, as indicated in Fig. 3.5. In the low carbon range of iron–carbon alloys, wrought and casting alloys are very similar, and, as heat treatment can be applied equally well in both cases, the final properties of the two alloy groups do not differ much. As the carbon content increases from medium to high carbon steels, it becomes increasingly difficult to control the fineness of the Fe_3C or cementite phase by the casting and heat-treatment processes alone, and consequently wider variations in structure and properties between cast and wrought alloys result. Finally, as the proportion of the Fe_3C phase increases in the hyper-eutectoid range, the alloys become difficult to work by plastic deformation, and the remaining alloys are essentially casting alloys. The first large family of higher carbon and silicon alloys are malleable irons, and the second grey irons. Alloys with a lower carbon and silicon content than malleable irons are also used for casting, but less frequently. As malleable and grey irons are typical as well as industrially very important casting alloys, their basic structural metallurgy will be summarised briefly.

The initial cast structure for producing malleable irons is known as 'white iron.' This forms during freezing and consists of Fe_γ, or primary austenite crystals, and a eutectic mixture of Fe_γ and Fe_3C, known as ledeburite (this cementite is defined as eutectic or primary cementite). On cooling through the eutectoid temperature all the Fe_γ breaks up into a fine lamellar eutectoid mixture, pearlite, containing Fe_α and Fe_3C (this cementite is defined as secondary or eutectoid cementite). The abundance of total Fe_3C in the structure in relation to a small amount of ductile Fe_α makes the whole structure hard as well as brittle. In the malleablising or annealing process, the alloy is heated back to the Fe_γ temperature range and held while graphite crystals grow from the primary Fe_3C, representing first stage graphitisation. The alloy is then cooled slowly through the eutectoid temperature range while the carbon dissolved in Fe_γ grows on to the graphite crystals formed during the first stage, thus giving second stage graphitisation. A typical annealing cycle, shown in Fig. 7.7a, allows several structural variations to be obtained at room temperature, the significance of which can be seen by considering the various structural changes occurring in the cycle. The initial structure should have no graphite flakes. This is controlled mainly by the carbon and silicon content for a given cooling rate which, in turn, depends mainly on the mould and casting design, particularly the ratio of surface area to volume of the casting. In addition, some of the subsidiary alloying elements such as sulphur, phosphorus, manganese, nitrogen, chromium and copper have an influence, as well as minor additions prior to pouring, e.g. bismuth to promote the formation of white structure, and aluminium and boron to control the subsequent kinetics of graphitising or malleablising processes. The rate of heating to malleablising temperature especially in the range of 350°C to 400°C affects the nucleation process in first stage graphitisation, Fig. 7.7*b*. This effect is also sensitive to prior additions of aluminium or boron to the liquid iron. The full mechanism of the effects of these elements on structural

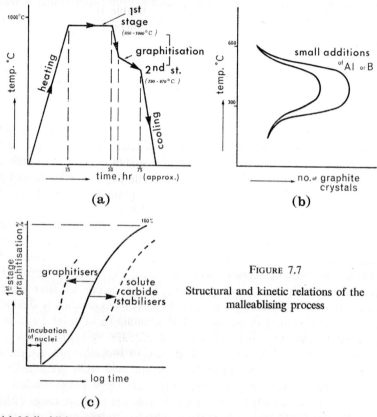

FIGURE 7.7

Structural and kinetic relations of the malleablising process

(*a*) Malleablising cycle.

(*b*) Sensitivity of first stage graphitisation nucleation to pretreatment temperature and composition.

(*c*) Kinetics of graphite growth during first stage of graphitisation as affected by time and alloying additions.

changes which affect the nucleation and growth of graphite crystals prior or during malleablising has not yet been fully explained.

At the start of the first stage graphitisation, microscopically discrete crystals of graphite begin to grow at the Fe_γ and Fe_3C interfaces, the kinetics of graphitisation being dependent on the annealing temperature and the carbide stability, in turn controlled by the overall alloy composition, Fig. 7.7*c*. The annealing cycle can be carried out in an oxidising atmosphere, so that some carbon is removed from the alloy. This process is called the whiteheart malleable process, due to the appearance of its fracture at room temperature. If the iron contains some free FeS in the structure, the graphite crystals grow in the shape of nodules but, if it contains free MnS instead, the graphite shape is more in the form of a flaky aggregate. If the annealing atmosphere is neutral, little or no carbon is lost, the fracture of malleable castings appears 'sooty' and hence the name blackheart malleable. After the completion of the first stage graphitisation, if the alloys are cooled slowly through the eutectoid temperature to allow all the carbon in solution in Fe_γ to grow on to the existing graphite crystals (second stage graphitisation), the final structure at room temperature consists of ductile Fe_α crystals with dispersed nodules or aggregates of graphite. If some carbon remains in solution in Fe_γ then some pearlite may form during the eutectoid transformation, giving various grades of pearlitic malleable irons. In addition to ferritic and pearlitic malleable irons, another family of malleable irons (heat-treated malleable) can be obtained by quenching from the Fe_γ range, followed by tempering.

Malleable irons are very good examples of alloys whose original cast structure is strongly influenced by small variations in composition and cooling rate. The kinetics of the solid state structural transformation too is strongly sensitive to the alloy composition, and also to the heating rates, cooling rates and composition of the atmosphere of the annealing cycle and, to a minor extent, to the original size and distribution of primary Fe_3C. The actual carbide composition in all industrial (FeSi)–C alloys should be designated as $(FeSiPCr)_xC_y$; Fe_3C is only used as an abbreviation. Malleable irons are therefore a good example of alloys whose cast structure influences equally strongly the process of manufacture and the properties of castings produced. The fact that the initial structure must be white restricts the size of such castings to relatively thin sections only, Fig. 7.5b. If the alloy composition were altered for casting thicker sections, then the subsequent malleablising process would be too long. In the process of malleablising, a distribution of graphite is achieved that is least injurious to the properties of the matrix. Finally by changing the structure of the matrix a wide range of properties of castings can be obtained.

Unlike the malleable irons, the properties of grey cast irons are such that these alloys are generally used as-cast , and only in special cases are they heat-treated. Consequently, the main problem is to obtain the desired structure directly through control of the solidification stage, resorting to heat treatment only when the desired properties cannot be obtained directly. At the end of solidification the structure of grey irons contains crystals of Fe_γ and graphite. The shape and distribution of graphite crystals are controlled by the alloy composition and by the crystallisation process, as discussed in the next paragraph. When the Fe_γ crystals transform at the eutectoid temperature, some free Fe_α or ferrite may result if the cooling rate is slow enough. The carbon originally dissolved in Fe_γ either deposits on the available crystals of graphite, or forms Fe_3C crystals which, in conjunction with the remaining Fe_α, forms the eutectoid phase mixture, pearlite. The final cast structure, at room temperature, therefore contains only ductile crystals of ferrite and brittle and weak flakes of graphite when cooled very slowly through the eutectoid range, whilst with an increasing cooling rate brittle but stronger constituent pearlite, at first gradually and then fully, replaces the ferrite during the eutectoid transformation. The main structural problems of grey irons are therefore to to control, firstly, the size, shape and distribution of the graphite crystals and, secondly, the proportions and distribution of pearlite and Fe_α.

Numerous metallurgical studies of crystallisation of graphite in cast irons have demonstrated that the variations in the occurrence of graphite are largely controlled by three main variables—the presence of alloying elements other than iron, carbon and silicon, the impurities in the alloy, and the cooling rate. The effect of the main alloying elements carbon, silicon and phosphorus on the amount of graphite formed at slow cooling rates, such as in sand moulds, can be expressed in terms of an empirical carbon equivalent % by the relation:

$$C.E.\% = C\% + \frac{Si\%}{3}$$

Since phosphorus is an element frequently present in cast iron and also affects it, the carbon equivalent may be modified into:

$$C.E.\% = C\% + \frac{Si\% + P\%}{3}$$

The C.E.% combines the effects of the main alloying elements carbon, silicon, and phosphorus, and shows that silicon and phosphorus have about one third the graphitising power of carbon. The C.E.% can also be directly related to the tendency to graphite formation during freezing, as affected by variations in cooling rates due to the differences in the cross-sectional areas of castings in sand moulds, as shown in Figs. 7.8a and b. The effect of the other alloying elements can also be expressed in terms of either the graphitising or the carbide stabilising power of carbon or silicon and plotted in C.E.% graphs. The main practical application of the C.E.% is in selecting an iron composition to ensure that graphite will form in the structure for a given casting cross-section in sand moulds.

Impurities may have a sufficiently strong effect on graphite formation, its size, shape, distribution and quantity to alter significantly the simple relationship between C.E.% and the cooling rate. The exact mechanism of this has not yet been established. For example, superheating the alloy promotes carbide stability, and many elements known as carbide stabilisers (Cr, V, Te, N) if present as impurities may have a similar effect. As all variations in impurities cannot be readily controlled in normal foundry practice, it is useful to inoculate cast irons prior to pouring; for example with 0·1% to 0·5% of Fe–Si alloy, or CaSi, or FeSiZr, FeSiSr or SiC or graphite. An inoculation treatment evens out the effects of unknown impurities on the solidification process and allows a more reliable prediction on graphitising power from

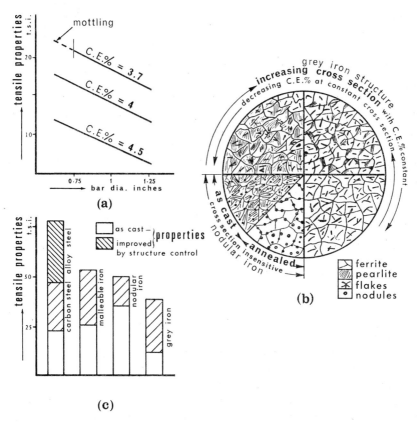

(a)

(b)

(c)

FIGURE 7.8 Tensile properties: structure relations in (FeSi)—C alloys

(*a*) Cross-sectional sensitivity of flake graphite irons.
(*b*) Structural variations in flake graphite irons with C.E.% and cross-section.
(*c*) Range of properties obtained by changing the structure of matrix, graphite and Fe_3C crystals.

C.E.% values. Similarly, inoculation promotes graphite formation, and thinner 'grey' casting sections can be obtained with the same C.E.%.

While the above relationships can be used to explain the problem of general structural control of grey irons in terms of the amount of graphite, there still remains the problem of the shape and distribution of the graphite in cast iron of a given C.E. %, and the effect of such structural variations on the properties of grey irons. Two main graphite crystal shapes occur; flakes and nodules, but several types of distribution of flake graphite in the structure of grey cast irons are also possible. Variations in properties of unalloyed grey irons are thus obtained either by distinct distributions of flake graphite crystals or by the nodular or spheroidal graphite shape, or by matrix structure variations in both cases. Undercooled, eutectiferous, or 'D' flake graphite, Fig. 6.8b, is formed with decreasing C.E. %, or by an increasing cooling rate when the cooling conditions approach closely the appearance of white eutectic. 'D' flake graphite is therefore common in the mottled structures of irons (containing both graphite and ledeburite). This form of graphite in the structure is unfavourable from the point of view of some mechanical properties and machinability of grey irons. A uniform distribution of flake graphite designated as 'A' graphite, is usually preferred, Fig. 6.8a.

If the amount of sulphur in an iron solidifying with an 'A' graphite is reduced below the value of approximately 0·001 %, the flake graphite crystals are replaced during crystallisation by nodular graphite or spherulites. At the same time, the nucleation behaviour of these graphite crystals is affected and an inoculation treatment is necessary, as with normal flake irons, to avoid ledeburite. Furthermore, for the same cooling rate, an iron with flake 'A' graphite may have only a small amount of pearlite, while the same nodular iron would have a fully pearlitic matrix after the removal of sulphur. A structure with a high pearlite content is hard and difficult to machine, for many industrial applications necessitating an annealing treatment to convert pearlitic into ferritic nodular irons. Thus in the end, nodular or S.G. iron is structurally very similar to a whiteheart malleable iron, and has similar mechanical properties, Table 7.1.

The properties of grey irons are determined by the two main constituents. The matrix has all the structural possibilities of steels, varying from ferritic to pearlitic structures, or all the combinations of Fe_α and Fe_3C obtainable by various heat treatments of the matrix. The dispersed phase, graphite, when in the shape of flakes, has an embrittling effect on the matrix, but with the nodular type the whole structure becomes more ductile. With flake graphite irons the strength properties increase with a decreasing C.E. %, or with a decreasing cross-section of the casting and a constant C.E. %, as the total volume of graphite in the structure decreases. By variations of the matrix structure and changes in graphite shape, size and distribution, a wide range of mechanical properties can be obtained in grey irons, Fig. 7.8c. It is also clear that there is little strength to be gained mechanically by changing the matrix of flake irons by heat treatment. Widely different and specific matrix properties (temperature or corrosion behaviour) can also be obtained by the addition of alloying elements, giving a whole new series of alloy grey or nodular irons. The same applies also to malleable irons.

7.7 SELECTION AND MAKING OF CASTING ALLOYS

Metallurgically, the term casting alloy as distinct from wrought alloy implies an alloy which has certain distinct and favourable casting properties: ease of melting and compositional control, good fluidity, relative freedom from tendency to segregation or to hot-shortness, readily controllable cast structure, and freedom from shrinkage cavities.

This explains why it is that most casting alloys used in the industry are structurally eutectiferous, as these alloys possess the best combination of the properties listed. Wrought alloys are generally of the solid solution type with the required amount of dispersed phases to control plasticity, strength or other properties. Castings are also made in these alloys, but foundry processes become more difficult. A good example of this kind are many steel casting alloys where compositions of both casting and wrought alloys are identical. On the other hand, flake graphite grey cast irons are typical casting alloys which are seldom, if ever, processed by metal working. Furthermore, grey cast irons are a good example of an alloy which combines good casting properties with a range of useful engineering properties (machinability, corrosion resistance, moderate strength and damping capacity), as well as an overall favourable production economy. The problem of industrial usage of casting alloys therefore requires joint consideration of metallurgical, engineering and production requirements.

Major casting alloy groups used in the foundry industry are based on iron, copper, nickel, aluminium, magnesium and zinc. In each case there are subgroups based on binary, ternary or more complex alloys. Making up of the composition of casting alloys based on the above elements follows the general principles discussed in this chapter. The type and quantity of the various elements used aim to obtain the main process and property requirements. Further additions may be made to control a specific casting property or a particular property in the finished casting. Thus in general the actual chemical composition of currently used alloys in industry is frequently a compromise between metallurgical, engineering and production requirements. However, many elements given in the standard specification of a casting alloy are there, not because of any beneficial effect to the alloy, but because their removal from the raw materials available would be uneconomic for the process as a whole. Economic factors and the availability of metals also limit the scope for development of entirely new 'tailor made' casting alloys. Considerable room however still remains for improvements in the composition control of structure of old-established alloys and for formulating directions in changes of composition to obtain new structures with distinct properties.

7.7.1   Selection of casting alloys

Before selecting a casting alloy the property requirements of the finished casting are first considered. In some cases a physical property, such as electrical or thermal conductivity, limits the choice to one or two alloys mainly based on copper. Similarly, damping capacity and acoustic behaviour determine the choice of bronzes or carbon steel for bell castings. On the other hand, mechanical property requirements can usually be met by several casting alloys, and here the casting properties, or production and economic requirements, may be the deciding factors. The selection of a casting alloy for general applications (unstressed components) is sometimes made on the basis of a specific property, but more often than not on purely economic grounds.

7.7.2   Cast, wrought and other types of alloy structure

Apart from the cast structure and different kinds of wrought structure that can be obtained from cast structure by subsequent working and heat treatment processes, solid objects in alloys can also be formed from metal powders or by deposition from the gaseous state or by electrolysis. If a given solid object is made in the same alloy but by any one of the different processes, then the resultant structure will be different to a smaller or larger extent in each instance. As in every manufacturing process, such as casting, a number of different kinds of structure can be obtained, it is clear that a major metallurgical problem in manufacturing is that of understanding the structural behaviour of alloys. The basic crystals making up the structure are the only feature common to all the different structures, whilst the main differences amongst them lie partly in the sizes, shapes and distribution of all the crystal phases present, and partly in the types, amounts and distributions of various imperfections and defects. As sizes, shapes and distributions of crystals can also be considered in relation to imperfections of various kinds, it is clear that the differences in all the structural conditions, and hence the properties of alloys, which can be encountered are due to variations in the occurrence of imperfections. To obtain optimum specific properties it is necessary to reduce to a minimum certain kinds of imperfections which affect the particular properties.

7.8   SUMMARY

The extent of the general application of an alloy in industry depends on the availability of raw materials, the technical feasibility and the economy of the manufacturing process, and the possibility of obtaining a wide range of product properties for different conditions of their application. Casting processes offer an intrinsic economy in the way solid shapes can be produced which require only minor dimensional adjustments for their application.

Applications of castings are therefore often controlled by the success of achieving the correct cast structure to meet the property requirements. In practice, this means not only controlling the structure of main crystals but more often than not controlling various types of imperfections and defects, particularly those associated with cavities and inclusions. In the field of founding in general, the craft of founding implies mastering the art of producing cast shapes, the engineering contribution to the industry lies in the means and methods which control the economy of casting processes, whilst the metallurgical contribution is largely that of controlling the cast structure and its properties.

FURTHER READING

SMALLMAN, R. E., *Modern Physical Metallurgy*, Butterworth, London, 1962.
DIETER, G. E., *Mechanical Metallurgy*, McGraw-Hill, New York and Maidenhead, 1961.
BRICKS, R. M. and PHILLIPS, A., *Structure and Properties of Alloys*, McGraw-Hill, 1952.
PECKNER, D., *The Strengthening of Metals*, Reinhold, New York, 1964.
MURPHY, A. J., *Non-ferrous Foundry Metallurgy*, Pergamon, Oxford, 1954.
HANSON, D., and PELL-WALPOLE, W. T., *Chill Cast Tin Bronzes*, Arnold, London, 1951.
BOYLES, A., *The Structure of Cast Iron*, American Society of Metals, 1947.
Grey Iron Castings Handbook, Grey Iron Founders Society, U.S.A., 1958.
PIWOWARSKY, E., *Hochwertiges Gusseisen*, Springer, 1961.
Malleable Iron Castings, Malleable Founders Society, U.S.A., 1960.
BRIGGS, C. W., *The Metallurgy of Steel Castings*, McGraw-Hill, 1946.

CHAPTER 8

MOULDS

8.1 INTRODUCTION

The principal casting processes and their main characteristics were introduced and briefly described in Chapter 1. It was then pointed out that all stages of a casting process, from the casting design at the start to the evaluation and testing of the properties of the cast product at the end, the various engineering and metallurgical problems overlap and are superimposed one upon the other. This is best shown in what is the central and focal production stage of founding, namely the construction and design of moulds for the casting process. The engineering sciences are concerned with problems of design of castings and moulds, with equipment and methods used for mould making and with the process operations. Metallurgy deals with the materials used, their properties and their behaviour during the process. In line with the general theme of this book, metallurgical phenomena encountered in the casting process will be considered and will be subdivided into four groups following the production sequence of the process, namely, the mould (Chapter 8), filling of the mould (Chapter 9), solidification and feeding of the casting (Chapter 10), and resultant properties and examination of the cast products (Chapter 11). The present chapter deals with the properties and behaviour of mould materials, and the discussion can conveniently follow the sequence of mould making in a foundry: (a) construction and design of moulds, (b) behaviour of moulds in the casting processes, (c) testing and control of properties of moulding materials and (d) metallurgical significance of moulds.

8.2 MOULD CONCEPTS, DEFINITIONS AND MATERIALS

The general requirements of an ideal mould are (a) to contain a cavity which is a faithful reproduction of the shape of the casting which has to be produced (b) to withstand the filling process and to extract the heat from the molten metal in such a way as to produce optimum properties in the casting, and (c) that it should be constructed and used in the most economical manner possible. The history of founding is largely the history of different solutions to the economic, engineering and material problems of moulds. One example may illustrate this point. Pressure die casting metal moulds ensure a very high dimensional accuracy of castings, have a long life, but also have severe material and economic limitations where complex designs, heavy castings, high temperatures and optimum properties of castings are concerned. Sand moulds, on the other hand, are almost free from some of these limitations, but give a comparatively low dimensional accuracy and last generally for only one casting.

This example indicates how mould materials can be divided into two groups: (a) metallic, also called chill, or permanent moulds, which can be used repeatedly, but still give a finite mould life, and (b) silica sand and other mineral moulds which are broken up to remove the casting (consumable moulds although some of the moulding mixtures are often reused, 8.9.4).

In engineering practice a metal component is first conceived as a drawing. A study of the design features, the properties of the casting required and of the economic factors determines whether the component in question should be made from a wrought product or from wrought and cast products, or should be made as a single casting. If the latter, it must be further decided whether a metal or a mineral mould casting process should be used. In many instances this is obvious, as, for example, the choice of pressure die casting for producing carburettors for motor cars. Conversely, either metal or sand moulds can be used to produce oil sump housing for car engines, and moreover this type of shape could also be produced as a pressing. The selection of a moulding material, and hence the choice of casting process, will be made clearer after examining the metallurgical questions arising with the moulding materials used at present in foundry practice. The remainder of this chapter is therefore devoted to moulding materials, with special reference to their behaviour as observed and recorded in current foundry practice.

8.3 METAL MOULDS AND PROCESSES

Before dealing with the metallurgy of metal moulds, the essential features of the casting processes where such moulds are used will be briefly outlined.

8.3.1 Ingots

Some general concepts of ingot casting were introduced in 1.3. Major divisions of batch moulds are: horizontal or vertical (position of longest axes), solid or water-cooled, single or split unit (for ingot removal) and parallel sides or tapered. Geometrical variations of size and design can be related to the working process used for ingots (type and extent of deformation and the type of products made) and to the properties of the alloy cast (casting and working properties). Different cooling methods are selected to meet ingot property requirements (mainly plasticity) consistent with optimum production characteristics (mould life, ingot yield and productivity). In a given process metallurgical control includes the following main factors: mould (temperature levels, time cycle and dressings), pouring conditions (rate, turbulence and metal distribution) and metal (pouring temperature and compositional quality).

The changeover from batch to semi- or continuous ingot casting processes results from a combination of metallurgical and production requirements. In simplest terms, the improvements are either in metallurgical quality (Chapter 11) or in overall production economy. On the whole, metallurgical benefits decrease and technical casting problems increase the higher the casting temperature of the alloys. The production factors (capital investment and productivity) favour high tonnage alloys, i.e. most wrought alloys in light metals, some copper base alloys and plain carbon steels.

Batch and continuous methods are often competitive. Engineeringwise, batch ingot equipment and operations are fairly simple in contrast to the often complex machines and controls of continuous casting processes.

The main subdivisions of the continuous casting process are: semi- or fully-continuous pouring (1.3.2 and Fig. 8.1b); open or closed moulds (Fig. 8.1a); vertical, inclined or horizontal ingot withdrawal; casting of primary ingot shapes or of semi-products (thin sections and small sizes); casting only or combined casting and working in one machine unit. Process selection in this field is therefore complex and some major factors are: semi-continuous processes are favoured when the production output is rather small balanced with the need for a variety of products in different alloys; closed and/or horizontal mould methods are more suitable for casting smaller size (cross-section) ingots; casting of semi-products is advantageous for simpler sections (strip or sheet); and combining casting and working into one is a productivity choice. Although any given continuous casting process has a large number of possible engineering design and operation variations, some process control characteristics are common to all the methods. Metallurgically, the major joint factor is heat transfer control, i.e. synchronising and balancing temperature gradients and their directions with the casting speed (metal and cooling system temperature and rate controls). For some alloys and cast shapes, heat transfer parameters to produce satisfactory ingots can vary within wide limits, while for others the process can be operated only under a restricted range of variations in casting conditions.

8.3.2   Gravity die casting

By definition (1.3.6) a shaped casting poured into a metal mould under the force of gravity is described as gravity or permanent mould casting, Fig. 8.2(2). (Some ingots which are used as-cast or machined and not subsequently worked, e.g. for bearings, are the simplest form of gravity die casting.) The general advantages of gravity die castings are to be found in their surface and dimensional properties (combined with their reproducibility) and relatively high production rates. However, mould life of all materials commonly

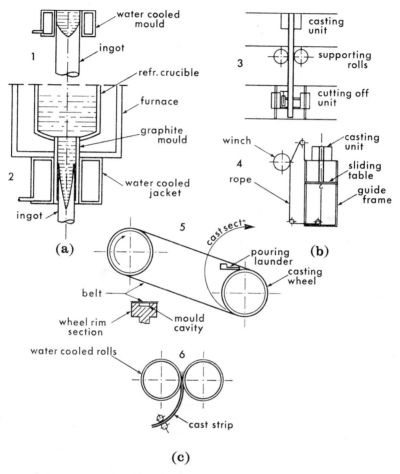

FIGURE 8.1 Some continuous casting methods

(a) (1) open and (2) closed pouring systems.
(b) (3) continuous and (4) semi-continuous devices.
(c) Casting of semi-products: (5) wheel method (Properzi) for strip ingots, and (6) roller method (Hazellet) for sheet.

available is rapidly reduced as the pouring temperature of the alloy increases. (This is also true of batch ingot moulds, even with their simple geometrical shape.) Furthermore, mould making and operational costs rapidly increase with the complexity of casting design or with the weight of the casting.

Finally, the complete filling of the mould details and solidification problems (particularly casting stresses) demand favourable casting properties in the alloys used. These factors explain why this process is practised on a very large scale with light and lower melting temperature alloys only and to a

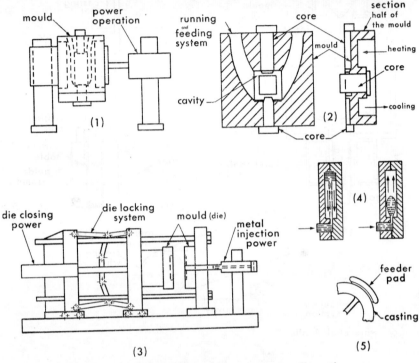

FIGURE 8.2 Elements of gravity and pressure die casting

(1) Engineering elements of gravity casting.
(2) Lay-out of a gravity mould.
(3) Engineering elements of pressure die casting.
(4) Flow phenomena in pressure die casting moulds.
(5) Feeding a pressure die casting.
Engineering elements include: power for opening and closing the mould, core operations and casting ejections; ancillary services, heating, cooling and mould dressing.

smaller extent with some copper base alloys and grey irons when simplicity of casting design, small weight and mould making costs are favourable.

The gravity process has a number of engineering variations. The moulds can be made in a precast shape in grey iron and finished off by machining or be fully machined out of solid blocks. Normally, the core components of moulds are made in wrought materials (steel) because of the severe conditions of their use. However, for complex internal shapes, mineral (consumable) cores have often to be used. The required mould temperature is determined by the metallurgical factors of mould filling and solidification conditions, as well as by operational factors, i.e. maintenance of mould temperature at an optimum casting time cycle (heating of the mould by the incoming metal and its subsequent cooling in order to repeat the process). The castings are normally poured under gravity, but in the low pressure process the mould is held above a sealed crucible containing the liquid metal which is forced upwards by the application of low gas pressure. Further possible engineering variations include mould opening and closing mechanisms and applications of power for casting operations and controls (semi- or full automation), Fig. 8.2(1).

8.3.3 Pressure die casting

Experience in pouring metals into metal moulds under the force of gravity demonstrates that the rapid rate of cooling and solidification makes it difficult or impossible for production and engineering purposes, to fill very thin sections of moulds (approx. <0·2 mm thick) or to reproduce the finer mould details (e.g. cast-in threads). These difficulties can be overcome by injecting the metal into moulds (dies) under pressure (of the order of 1 to 10 ton/in²). The pressure die casting process is thus the next natural stage of the development of the gravity process. However, the application of high mould filling pressures require important process modifications, particularly in the choice of materials and in the making of dies and the design of machines for their locking and unlocking, and filling under pressure, Fig. 8.2(3). To prevent erosion of the die face, the cast iron used in the gravity process has to be replaced with forged alloy steel for the working faces of the die, whilst less expensive steels are used for the remainder of the die construction. Consequently, all parts of a die are made by machining processes. To operate the pressure casting process it is necessary to design die and core components not only for easy locking, unlocking and casting removal, but also for withstanding high pressures during the filling. A pressure die casting unit can therefore be a complex engineering machine depending on die locking systems, methods of metal injection into the die, process control devices and the degree of automation.

The metal can be injected into the die by one of two main methods. In one method, a goose-neck shaped channel joins the die at one end to a cylinder located in the liquid metal holder (crucible), and the metal is forced into the die by applying pressure in the cylinder (hot chamber die casting process). In the other, the required amount of liquid metal is poured into a tube directly joined to the die, and the metal is forced into the die by applying pressure on to a piston located in the tube (cold chamber process), Fig. 8.2(3). The term 'cold' implies here that the metal injection temperature can be nearer its freezing temperature than in the hot chamber process. Material properties, temperatures required and engineering problems of injection lead to a general application of the hot chamber method to zinc and the cold chamber process to light or copper-base alloys. In a given process, the operating variables—casting cycle times (injection, dwelling, ejection and locking), metal and die temperatures, die dressing and injection pressure variations—are mainly determined experimentally.

On the engineering side, the pressure die casting process is thus characterised by relatively high initial cost of plant and equipment and die making. The process is therefore more economic when a large number of the same casting is required (a typical minimum number is of the order of 3000 to 5000). Fast production rates, high dimensional tolerances and small finishing costs are the major advantages, and restrictions in the choice of casting alloys, design complexity and casting weight are the main limitations of the process. Metallurgically, injection and temperature problems restrict the process to the lower melting temperature range alloys (mainly zinc and light alloy base), and the freezing conditions (casting stresses) limit the process to those alloys with unusually favourable casting properties (relatively few zinc and light alloys).

8.3.4   Centrifugal casting

The need for a centrifugal method of filling moulds arises partly from some of the limitations of gravity and pressure die casting processes and partly from some inherent advantages of centrifugal force applications. For example, both the making of the die and its use restrict the metal mould processes to relatively small weights of casting (approximately 1 lb to 200 cwt in the gravity process and from 1 oz to 50 lb in the pressure process). However, for simple shapes such as discs, short cylinders or long pipes, this weight range can be increased to 10 cwt or more, and the application of the process can be extended to copper alloys, cast irons or steels, whilst retaining the application of metal moulds Fig. 8.3a(2). Furthermore, if the metal mould is replaced with a mineral base mould material, the process can be applied to small or comparatively heavy castings (from 0·1 oz to 10 cwt) and without any restriction of casting design (shaped castings), Fig. 1.4c. The intrinsic value of the application of centrifugal force is that it provides a very simple and inexpensive method of filling mould details as well as exerting a pressure on the liquid metal during solidification, thus achieving a good measure of control of casting soundness (Chapter 10). With metal moulds, therefore, centrifugal casting is generally used for casting simple shapes; with mineral moulds any shapes can be cast. The choice of alloys is very wide, but not entirely unrestricted, as clearly some methods of mould filling, such as that in the casting of pipes, Fig. 8.3a(2), could lead to heavy oxidation of some liquid metals such as light metals, and other alloys are hot short to withstand the centrifugal stresses.

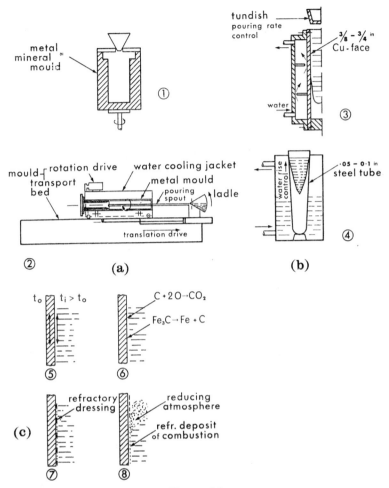

FIGURE 8.3

(a) Centrifugal casting: (1) solid metal or mineral mould, vertical or horizontal axes, for short pipes and cylinder liners, (2) water-cooled horizontal metal mould for long pipes.

(b) Water-cooling: (3) Cu faces and (4) thin steel tubes in batch ingot processes.

(c) Physical and chemical phenomena at metal—mould interface: (5) expansion, (6) surface reactions with grey iron moulds, (7) surface protection, (8) surface and molten metal protection.

The engineering problem of centrifugal casting is mainly that of designing a machine in which the centrifugal force can be best adopted for mould filling for a given casting shape. For long pipes, this is conveniently achieved by supplying liquid metal at a given rate, the mould rotating and moving away at a given speed, Fig. 8.3a(2). For short cylinders, no translation movement is necessary, and the rotation axes can be horizontal or vertical. For mineral moulds, the most convenient arrangement is that of spinning a vertical axis around which a number of moulds can be mounted (centrifuging), the liquid metal being supplied via the common central runner. A symmetrical casting can be spun around its own main axes. The major process variable of centri-fugal casting is the rotation speed. The optimum speed is that which achieves successful mould filling without the magnitude of the centrifugal force reaching values that cause hot tearing of the casting (the usual range is 200–1000 rev/min.). As centrifugal casting is essentially a slow process of production, its general applications are restricted to symmetrical solids or small shapes (jewellery and dental applications) or products of such a shape that metallurgical properties cannot be readily achieved by normal gravity pouring processes (castings for high temperature applications).

8.4 METALLURGY OF METAL MOULDS

8.4.1 General characteristics

The main production characteristics of metal moulds are their permanence or relatively long life (which varies depending on the details of the casting process), higher dimensional accuracy of the mould cavity—and hence of the castings—than that of most mineral moulds, and the possibility of a high production rate. Nevertheless, the costs of moulds and casting machines are high, and to make the use of metal moulds economically acceptable a large number of castings of the same design is usually required.

8.4.2 Mould materials

Two main considerations arise at the mould making stage: (a) the selec-tion of the metal composition from which the mould is to be made and (b) the question whether the mould material should be a cast or wrought alloy. In most instances these two problems are considered together. Castings are generally the more economic construction material and cast metals are used for mould construction wherever their performance is found satisfactory. When the design of the mould so requires, or the nature of the casting process is more exacting, wrought metals are used for mould making. Cast irons of various compositions are the most widely employed materials for solid batch ingot moulds and for gravity die moulds, while copper is used for water-cooled ingot moulds, and forged steel for pressure casting dies and for centrifugal metal moulds. The exact alloy specifications for mould materials are to be found in the textbooks recommended for further reading. The tradition of foundry practice has established a wide usage of certain materials for moulds. Factors to be considered, therefore, mainly concern the exact control of the metallurgical properties and quality of these materials, and the question of their replacement by other materials is less frequent. This discussion will, therefore, be confined to these factors.

8.4.3 Design of moulds

The general demand of a mould design is that it should extract heat at the required rate from the molten metal. The problem of mould design in practice is largely solved empirically, owing to the complex nature of heat transfer phenomena in moulds. Two main principles are applied: (a) when solid moulds extract heat too slowly, water cooling is generally used to increase the rate of heat transfer, and (b) the direction of heat flow is also controlled, so that castings solidify in the suitable directions.

More specific aspects of mould design concern some engineering and metallurgical functions of moulds. These are: the flow system of channels for introducing the liquid metal into the moulds, the feeding system for controlling the volumetric contraction of the casting, and finally designing the mould so that its component parts can be readily assembled for pouring the liquid metal (mould closing) and subsequently opened for removing the casting (stripping).

Examples of these principles applied in casting practice are illustrated in Figs. 8.2 and 8.3. The principle of water cooling is essential for continuous ingot casting, but it is also widely used for batch ingot casting. However, for steel ingots, the problems of the life of copper mould walls and the design of heavy size moulds make water cooling impracticable for batch moulds. On the other hand, water cooling is frequently used for gravity, pressure and centrifugal moulds. An example of die shape, assembly of die parts and application of directional solidification principles is illustrated with a gravity mould in Fig. 8.2. The mould contour, its wall thickness and cooling are arranged so as to achieve directional solidification of the casting, as well as rigidity and permanence of mould shape. An example of the flow and feeding problem is also illustrated in Fig. 8.2 with a pressure die casting. The flow pattern, including turbulence, can vary as shown, and air entrapment and cooling problems have to be resolved in order to achieve the required quality in the casting produced.

Mould design is a field where engineering and metallurgical considerations are strongly interwoven. It is the stage that largely determines the failure or success of the casting process, and hence it calls for the closest co-operation between the engineer and the metallurgist. Flow and feeding phenomena are more fully discussed in Chapters 9 and 10 respectively.

8.4.4 Mould behaviour and failure

The behaviour of moulds during the casting process explains largely the choice of materials for metal mould making. The working face of the mould receiving the molten metal is heated to a temperature higher than the initial mould temperature, creating a temperature gradient in the mould wall. In most batch casting processes the mould maintains such a temperature gradient as the casting cycle is repeated. The following factors have therefore to be considered for batch moulds in general: (a) the maximum and minimum temperature of the mould face in a cycle, (b) the steepness of temperature gradients extending from the mould face and (c) the frequency of the casting cycle.

As a consequence of heating to, or fluctuation of, mould face temperature in a casting cycle, the face repeatedly expands in relation to the cooler parts of the mould. Microstructural changes in the mould material may also occur as a result of prolonged heating at higher temperatures. The expansion of the mould face may be partly elastic and partly plastic, which may lead to mould warping. In addition to warping, the mould face may fail by thermal fatigue cracking. Microstructural changes in the mould face are either due to the chemical reactions which occur between the various gases present and the mould material or, with some alloys, slow structural phase changes may occur. With pressure die castings, flow impingement by the liquid metal stream and cavitation phenomena add to the causes of mould surface failures. During continuous casting of ingots, the mould surface wears off by friction from the moving ingot, and this rather than the temperature effects is one of the main causes of mould failure.

The metallurgical problem of metal mould selection is therefore that of choosing such a structural condition for a given mould material as will best withstand the conditions leading to mould failure. In addition, mould life is controlled by the process variables, as discussed in the next paragraph.

8.4.5 Process variables

There are very few instances where it is possible to expose a clean metal mould face to the incoming liquid metal. Apart from the danger of the liquid metal adhering to the mould face, the surface of the casting produced would be rough owing to the sudden solidification of the turbulent metal stream. Generally, therefore, the working face of a mould is coated with a thin film of a suitable mould dressing. Mould dressings for metal moulds fall broadly into two groups: non-volatile or refractory (inert) fine powders made of inorganic substances mixed with suitable binders (8.6), and volatile or flaming dressings based on one or more organic substances. In principle, a refractory dressing protects the mould face only, while the volatile dressing aims at protecting the mould face and the incoming metal from excessive oxidation by providing a partially reducing atmosphere inside the mould during pouring. The type, amount and method of application of mould dressings are determined empirically.

Mould temperature control has the following main objectives:

(*a*) to ensure safe usage of the mould, as damp moulds may lead to accidents through moisture blowing (gravity and ingot batch moulds are therefore warmed up prior to pouring to not less than about 80°C).

(*b*) to obtain a smooth surface on the casting; for example, in gravity, pressure and centrifugal casting the optimum mould temperatures may range from 200 to 350°C.

(*c*) to promote optimum direction and rate of solidification and produce fully fed, sound castings free from cracks.

(*d*) to obtain the longest possible mould life.

The detailed aspects of temperature control, such as optimum level, methods of maintaining optimum level, initial methods of heating and timing of the stripping of the casting, are determined experimentally.

8.5 SAND OR MINERAL MOULDS

While metal moulds, as we know them today, are a relatively recent development, sand or mineral moulds have been used since prehistoric times. The basic principle of the early moulding process has hardly altered, but there exists now a large number of distinct processes that differ from each other in practical detail. With modern advances in the engineering and metallurgical sciences, it is advantageous to consider sand moulding in terms of fundamental concepts rather than deal with various processes separately. However, before discussing such basic ideas and concepts, the main sand moulding processes will be briefly outlined.

8.5.1 Sand moulding

Sand moulding still accounts for the largest tonnage of production of shaped castings. One of the reasons for this is the natural abundance of moulding mixtures containing a mineral, such as quartz, of fine particle size, which occurs frequently premixed with suitable clay (8%–15%). The mixture is plastic and readily takes the shape of a pattern, provided the moisture content of the mixture is adjusted to approximately 5%–10%. A typical engineering casting of simple design is produced in the following stages: from the blueprint or drawing a pattern is made which is then rammed up in the moulding mixture, using moulding boxes to support the sides of the mould, the pattern is removed, cores are placed in position, the two or more parts of the mould, made in separate boxes, are closed up, and the metal is poured into the mould cavity, Fig. 8.4. The mould is as a rule made in two halves, top and bottom (cope and drag), and the number of cores may vary from none to several dozen with complex casting designs. However, three or four part moulding box assembly may sometimes be necessary. The whole problem of designing and sectioning a pattern into the required number of parts, arranging the joint lines and designing the core boxes so that the moulding

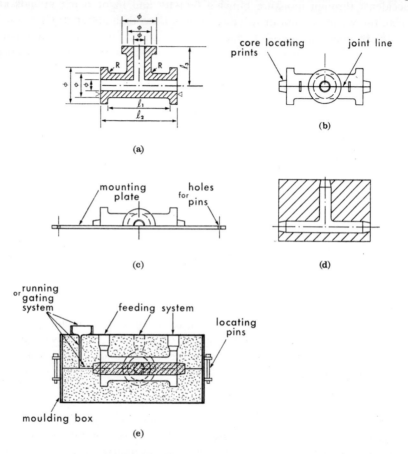

(a)

(b)

(c)

(d)

(e)

FIGURE 8.4 Elements of sand moulding

(a) Fully annotated drawing (the blue print).
(b) 'Loose' pattern for hand moulding (includes shrinkage allowances), split into two along the joint line for ease of moulding.
(c) Half of the pattern mounted on a plate for machine moulding.
(d) Section through the core box (for either hand or machine moulding).
(e) The two-part mould made, cored and closed.

process meets all the engineering, metallurgical and production requirements of moulding, is the very heart of foundry skill and/or engineering.

The main stages of the sand casting process—designing the castings, making patterns, ramming and closing the moulds, making and setting the cores, pouring the metal and cleaning or fettling the castings—can be carried out in a variety of ways to conform with different technical requirements, and this gives rise to many types of foundries. Pattern making is a large industry of its own, using a variety of techniques and materials. An example of splitting a pattern (joint line) for the purpose of moulding and mounting for production requirements is shown in Fig. 8.4.

Sand moulds can be made by hand or by using a wide variety of moulding aids or machines; these are made by the foundry equipment industry. The engineering problems of pattern making, of the design of foundry equipment and of setting up a production foundry process are a distinct part of founding, treated in textbooks of foundry engineering (Chapter 1). Some of the variants of sand moulding were introduced in Chapter 1, whilst more detailed characteristics of moulding mixtures are summarised in Table 8.1.

8.5.2 Core moulding

In order to produce internal surfaces in most castings (e.g. radiators for central heating), solid shapes or cores are made of the required moulding mixture and placed inside the sand moulds. Some typical core mixtures are summarised in Table 8.1. After pouring the casting, the core is removed from inside the casting, leaving the desired internal shape, Fig. 8.4. Core making therefore requires, in principle, similar considerations to those described for moulds, a core box in core making corresponding functionally to the pattern in mould making. The design of cores, their properties in casting (high initial strength to withstand the pressure, followed by collapsibility to allow free contraction of the solidified castings) and the need for their easy removal from castings (8.8.4) demand different moulding mixtures for cores from those used for moulds. This in turn leads to different types of equipment or core making, and often to a separate floor area in the foundry for this purpose, or the core shop.

8.5.3 Carbon dioxide, CO_2, process

This process can be used for making both moulds and cores, and differs from normal sand and core moulding essentially in the nature of the binder used to bond the particles of the mineral aggregate, Table 8.1. After ramming the mould or filling the core box, the weak silicate bond is hardened by passing through CO_2 gas for a short time (up to 1 min.), following which the mould or core is strong enough for immediate use in mould assembly and pouring. As a result of this special hardening technique, some marginal modifications in foundry equipment used for mould and core making have to be introduced. The bonding problem of this process is discussed in 8.6.

8.5.4 Shell moulding process

For making a normal sand mould, the whole pattern or a part of it is placed on a board within a moulding box, and the moulding mixture is then rammed up on the pattern. An alternative method is to cover the pattern with a suitable moulding mixture, which hardens rapidly to allow separation of a thin shell (0·2 to 0·5 in. thick) to obtain one half or part of the mould. The second half or outer part is made in a similar way, and the shells are glued or mechanically held together to provide a mould cavity for receiving the liquid metal. To make this process technically feasible for casting production, the shell must be strong enough to be lifted off the pattern, and at the present time this is mainly achieved by using a special type of binder in the moulding mixture, in combination with a heated pattern, as discussed subsequently in 8.6. Clearly this process can be used for both mould and core making, but requires major modification in pattern design and engineering equipment as distinct from the processes of normal sand-clay founding. Unlike sand or CO_2 moulding, the shell moulding process is restricted to relatively small size castings (up to 2 to 3 cwt), mainly owing to considerations of equipment and production costs.

8.5.5 'Lost' wax (precision or investment) moulding process

In both sand and shell moulding, two or more parts of a mould are made separately and then closed together to form a mould cavity, since the pattern has to be removed from the rammed or compacted mould. Another method of producing a mould cavity is to fully enclose a pattern in the moulding mixture, then remove it in such a way as not to destroy or damage the mould. Unlike sand or shell moulding, where a pattern can be used repeatedly to produce a large number of moulds in succession, the removal of a fully enclosed pattern from the mould requires the 'destruction' of the pattern each time a mould is made (consumable pattern). This can be achieved in several ways, for example, by using wax, frozen mercury or an organic compound like polystyrene. The oldest of these is the wax method: hence the name 'lost' wax process. The term 'precision' implies a high dimensional accuracy of pattern and of casting reproduction, and the term 'investment' describes the stage of covering the pattern with the moulding mixture, or investing the pattern. The various stages of the process are self-explanatory in Fig. 8.5. Apart from differences in pattern between the lost wax and other sand moulding processes, further differences arise in the moulding mixtures used, Table 8.1, and in other aspects of design and equipment process engineering. Precision casting, although largely restricted to special applications of casting design or to special (mainly high-temperature) alloys, is an important industrial process. It is also used for art casting, details of which are discussed in the text for further reading listed at the end of this chapter.

8.5.6 Multiplication of castings

The process of making castings has been discussed so far on the basis of obtaining a mould cavity of a given pattern. With relatively small size castings, and using any one of the different moulding processes, it is frequently advantageous from a production point of view to make a single mould containing several cavities of the same or similar castings, joined by

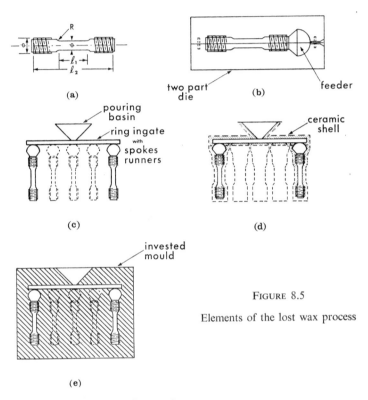

FIGURE 8.5

Elements of the lost wax process

(*a*) Blue print of a small tensile test piece.

(*b*) Mould or die (usually in metal) for injection of the wax at low pressure (approx. 50–100 lb/in²).

(*c*) Assembly of wax test pieces, first coat (dressing) applied by immersion.

(*d*) Ceramic shell mould made by alternate immersion into liquid mineral binder mixture and stuccoing with dry mineral over fluidised bed, then drying out and heat-shock melting out the wax.

(*e*) Solid mould obtained by investing the precoated wax assembly with liquid mineral binder mixture, vibrating, drying out and slow heating to 900°C to 1000°C. Moulds usually poured preheated to 500°C to 1000°C.

channels for pouring simultaneously. In this way a more rapid rate of production is achieved. Conversely, parts of a given casting can be moulded and poured separately, and the complete design is obtained by welding together the cast component parts (assembly method).

8.5.7 Résumé of mineral mould casting processes

From the above brief outline of the five major methods of mineral mixture moulding (others are indicated in Table 8.1), it is apparent that a feature common to all of them is the use of some specific moulding mixture for mould making; the differences are mainly to be found on the engineering side. A moulding mixture is made from a mineral, defined as a base or an aggregate, and from a binder, other additions being made in special cases. The metallurgical properties of such mixtures, and of the completed mould, can therefore be related to specific properties of the components used, and to the methods of compacting or ramming the mixture into the mould.

From the engineering and production point of view, mineral mould processes, in contrast to metal mould processes, require a new mould for each casting poured. In most processes (apart from the lost wax) mould construction involves one or more mould parts, and the dimensional accuracy of the castings produced depends on the accuracy of these parts as well as on the accurate assembling of the whole mould. Furthermore, a mineral mould may deform under the effect of liquid metal pressure, and moulds for heavier castings often have to be mechanically reinforced. In general, therefore, mineral mould castings will have lower dimensional accuracy than the corresponding metal mould castings. The rate of production of moulds depends on the size of the castings and the extent of the mechanical aids used, and can vary from about 300 moulds per hour for small castings to 1 mould per week for heavy castings (more than 2 tons approx.). A major characteristic of mineral mould processes is their flexibility and wide range of applicability, i.e. for a single casting or for a large number, for castings of simple or of complex design, for low or high temperature alloys, and for castings of wide variation in size. In addition, most moulding mixtures are cheap and readily available, and rates of production can be adjusted to suit the needs. The advantages of specific moulding processes can often be exploited in a single foundry. All these factors explain why mineral moulding processes are not only the oldest of all foundry processes, but also represent the basic foundry industry at the present time. Metallurgically, minerals used for moulds are poor conductors of heat, which leads to slow cooling and solidification of castings. This makes it difficult, particularly with some alloys, to achieve an optimum level of structure and soundness in the casting.

8.6 MINERAL MOULDING MIXTURES

8.6.1 The base

In principle, many kinds of mineral could be used for making a moulding mixture. The major requirements are determined by process, mould and economic considerations.

Foundry experience has shown that, for a given moulding process, the base aggregate should be within a given range of particle size, but complete control of particle size distribution and shape, though in some cases desirable,

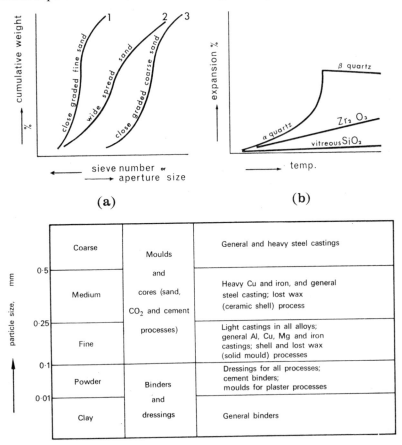

(a) Size grading of mineral particles used for moulding.
(b) Expansion behaviour of some minerals used for moulding.
(c) Particle sizes and typical applications for binders, dressings and mould materials.

is more difficult to obtain. Particle size varies in practice from 0·01 to 1 mm dia., and the size variation is normally shown in terms of either linear or area units, or is expressed simply in terms of standard sieve numbers. The particle size distribution of an aggregate can be plotted as shown in Fig. 8.6a. A smooth mould, and hence a smooth casting surface, are obtained by selection of a fine aggregate for the moulding mixture or, when this is not possible, by applying a fine aggregate to the face of the mould only as a dressing. In general, finer aggregates can be used for investment and shell moulding, whilst for sand moulding the aggregate size is largely determined by the properties of the alloy being cast, as discussed in 8.8.

Practically the only mineral occurring freely in nature in the required particle size range, and in sufficient abundance all over the world, is silica or quartz (SiO_2). Other minerals, oxides or silicates, zircon, olivine, chromite and sillimanite, have sometimes to be crushed and sieved, or even prefired; this is necessary with alumina or aluminous silicates (calcined aluminium silicate). This explains the preponderant use of silica sand for founding. This sand either occurs premixed with clay as a natural moulding mixture, or it is relatively pure and special clay or some other binder is added to make a synthetic sand mixture for various moulding processes. In the lost wax process, silica sand, largely because of its expansion behaviour, Fig. 8.6b, is not as satisfactory as some other minerals.

Some of the properties of the base mineral affect the quality of the casting produced (mainly the surface and dimensional accuracy). Apart from the smoothness of the casting surface, which is controlled by the mineral particle size, its quality is influenced by metal-mould surface reactions, metal penetration and fusion (8.10.1) and mould surface failure such as cracking. The dimensional stability of the mould face is affected by the volume changes due to heating and expansion of the base mineral. Additional properties required by the base aggregate are therefore (a) refractoriness, which is mainly controlled for a given mineral by its melting temperature and the presence of impurities, and (b) the absence in the mineral of structural changes accompanied by volume changes. For casting metals at higher temperatures ($>1300°C$) refractoriness of the base becomes more important, and when this cannot be economically achieved, refractory mould dressings can be used as an alternative method of improvement. Silica (quartz) undergoes a reversible allotropic transformation at 560–580°C ($\alpha \rightarrow \beta$ change), Fig. 8.6b, accompanied by volumetric expansion. The use of other minerals in place of silica is infrequent in sand moulding practice, on economic grounds, and the expansion behaviour of the silica, where it leads to casting defects, is generally counteracted by a closer selection of the particle size and distribution of the silica base, and by special additions to the sand, such as organic additives, which increase the hot plasticity of the mould face and prevent it from cracking (scabbing). Zircon and chromite are more and more replacing silica in steel foundries.

8.6.2   The binder

The choice of a moulding mineral base is a compromise between the frequently conflicting demands of mould quality and process economy, and the same is true for the binder. A base and a binder can occur naturally mixed together, for example in geological formations of silica-clay sands, or as a single chemical compound, such as Plaster of Paris. The industrial development of casting processes has tended towards mixing a selected base with a selected binder. The most important combinations used in foundry practice are: mineral with clay, such as natural or synthetic sands in sand founding; mineral with organic binders, such as core and shell moulding mixtures; and mineral with inorganic binder, such as sodium silicate, ethyl silicate and cement for moulding in CO_2, lost wax and cement respectively. Each of these binders contributes specific properties to the moulding mixture, but before discussing this, the properties of the main binders will be briefly considered.

Clay is a general term for a group of minerals, aluminous silicates, which occur freely in nature. They vary in their crystal structure, physical constitution and chemical composition, and there are thus several distinct types. Some of the structural and chemical features of clays are shown in Fig. 8.7. From the point of view of making a moulding mixture, the more important properties of clays are those of thin-layered structure and the ability to adsorb metal ions (base exchange) and water. Clay layers or micelles are very thin (15–30Å) and readily adsorb metal ions (Na^+ or Ca^{++}); thus water is also adsorbed on these ions. The adsorbed water joins one clay particle to another, and clay platelets through water to the surfaces of the mineral base, by molecular forces or polar bonds. This clay-water-silica bond is the basis of the green strength of a sand-clay moulding mixture, Fig. 8.7c. Different clays achieve optimum bonding at different clay and water contents (up to 3% to 10% of each) in the moulding mixtures, depending on their structure and constitution. Bentonite clays appear to be the most favourable in this respect, requiring the least amount of clay and water, about 3% to 5% by weight, for a given green bonding strength of the moulding mixture. The bonding properties of clays are thus dependent on the type of clay, and on the adsorbed ions and moisture content of the mixture. On heating above 100°C, the clay loses some of the adsorbed water, and the green strength of the bonded aggregate increases to that termed the dry strength. At a range of temperatures around 700°C the remainder of the water held in the clay is lost, and the dry bond further increases. However, this water cannot readily be retaken by the clay, which changes structurally and becomes dead, while the moulding mixture permanently loses the plasticity which is essential for moulding. On the other hand, the loss of adsorbed moisture only can be readily compensated for, and the bulk of the moulding sand can be re-used after knocking out the castings by reconditioning the mixture (8.9.4).

Organic binders can be grouped into natural products such as oils, starch and molasses, and synthetic products such as resins. The bonding action of an organic binder with the mineral bases can be achieved in several ways. A binder can be prepared as a viscous liquid, which holds particles of the mineral

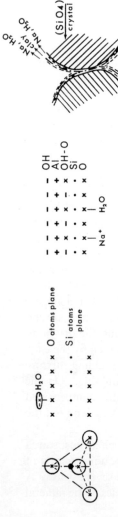

(SiO₄)—— tetrahedron Si–O covalent and ionic bonds H₂O–O polar (molecular) bond (surface adsorption)

Al(OH)₃–gibbsite hexagons Al–(OH) ionic bond

(a)

Kaolinite clay: alternate layers of (SiO₄)—— tetrahedra and gibbsite; at surfaces H₂O and Na⁺ ions adsorbed

Montmorillonite (bentonite) clays: silica–gibbsite–silica; Na (base exchange) and H₂O are adsorbed between clay layers more readily than with kaolinite

(b)

Bonding: surface of quartz crystals—moisture, base exchange ions—clay platelets—moisture, base exchange ions—quartz surface. Plasticity: slip along water/base exchange layers

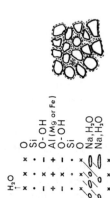

Bonding of sand-clay compacts: area of contact, clay film thickness, water/base exchange layer thickness

(c)

FIGURE 8.7

Structure and bonding of clays

(a) Structural units: silicate tetrahedron and gibbsite hexagons.
(b) Building up of various clays from structural units.
(c) Bonding and plasticity of mineral base and clay binders.

together by viscous forces sufficiently strong to enable a mould or core to be made by providing the green strength to the rammed mixture. The application of heat up to 150–250°C causes the binder to undergo structural changes accompanied by a large increase in strength (the baked strength). This is typical of the linseed oil or starch type of binder. The structural hardening process can also be achieved by chemical additions, or accelerators, to the binder-mineral mixture, giving a variety of air or room temperature hardening organic binders, as for example furane resins. Another possibility is to use a synthetic resin as a solid binder which, when mixed with the mineral aggregate, gives no green strength but will harden on a hot pattern, as in shell moulding and shell core making, or in a hot core box. Unlike the clay bond, which gives sufficient strength for the mould or core to be used immediately for many foundry applications, most organic binders have only a weak green strength, or none at all, so that structural hardening of the bond is necessary. Another way in which they differ from clay binders is that the binders break down and lose their bond when the temperature rises above 450–600°C. This property of organic binders, termed collapsability, is important for core removal from inside the casting after the process of solidification. These aspects of organic binders are further discussed in 8.8. The usual quantity of a single organic binder, or a mixture of several, added to the mineral base is 3% to 5% by weight. Water is sometimes one of the components of an organic binder mixture, for controlling either the viscosity or the bonding behaviour of the binder.

Inorganic binders are essentially silicates—sodium silicate or water glass, ethyl silicate, and calcium aluminous silicate or cement. As with organic binders, the green strength obtainable by mixing such binders with the mineral base is very weak, and a structural change in the binder is necessary to produce the required bond strength. With sodium silicate this can be achieved by passing CO_2 for up to 1 minute through the prepared core or mould rammed in a mixture containing 3% to 5% by weight of silicate binder. An alternative method is to apply a low pressure (approx. 1 to 5 mm Hg) and to remove some water from the silicate. Either of these hardening methods leads to structural changes in the silicate, and in the case of CO_2 application there are some chemical changes, which are not yet fully understood. In both methods a large increase in the bonding strength occurs. Bonding with ethyl silicate is achieved by hydrolysis, whereby the silica of the binder is deposited on the grains of the mineral base. After coating or investing the pattern with the premixed semi-liquid base and binder mixture, and after completion of hydrolysis, the excess water and alcohol are evaporated by heating, first slowly, and then at 900–1000°C. Hardening of a cement binder is also a slow hydrolysis process, but the reaction is achieved at room temperature. Some typical binders and their bonding action in moulding mixtures are summarised in Table 8.1.

TABLE 8.1 Typical moulding mixtures

Moulds and cores can be made in a single mixture or different mixtures for each. Simple binders are given, but in practice combinations of binders can be used. Other specific additions for moulding or alloy requirements are discussed in the text or in the further reading.

Process	Base aggregate		Binder			Bonding mechanism	Typical methods and applications
	Chemical nature	Particle size range, mm	Chemical nature	Weight %	Water %		
Green sand (synthetic, clay added to the base)	minerals: quartz, (SiO_2) olivine, (Mg_2SiO_4) zircon, $(ZrSiO_4)$	coarse 0·35–0·5 medium 0·15–0·35 fine 0·1–0·15	clays: montmorillonite $(OH)_4Al_4Si_8O_{20}$ $n\,H_2O$	2–3 3–5 5–7	2–2·5 2–4 3–5·5	Clay particles held together by ionic forces provide a continuous film around the particles of the base	Mixture of the base-clay and water, milled; very wide use for moulds for small and medium weight castings in all alloys
Green sand (natural mixture of base and clay)	mineral: SiO_2	medium to fine	clays: illite or kaolinite	10–15	7–9	Similar structure but different composition from montmorillonite, less plastic, more clay and water required for similar strength	Milled and used as above, but controls more difficult, hence practice localised to the areas of sand occurrence
Dry sand (either natural or synthetic)	as with the green sands					Removal of H_2O (150–250°C for 24 to 72 hrs) reduces weak H_2O links in the clay, raising the strength	Green sand moulds, skin or more fully dried (mould weight factor), for medium and large moulds or cores
Loam	as with synthetic or natural sand			5–7 10–15	15–25	As with clays, but supported by organic natural fibres (straw, horsehair)	Mainly for strickling moulds or cores of large regular shapes slowly dried (dry loam)

Table 8.1 (*Contd.*)

Core sand	minerals as with the green synthetic sands	carbohydrates starch, dextrin, cereals	2–4	1–4	Drying out and physicochemical changes in binder film, dissolved in water, sufficient green strength for handling	Hardening by heating 1–4 hrs. at 150–180°C (core size), small and medium cores (or moulds)
		natural oils: linseed, fish, soya	1–2·5		Oil film on the base particles polymerises in presence of oxygen on heating or by chemical acceleration at room temperature	1–4 hrs at 200–250°C, for light and medium cores (or moulds), or air setting 4–24 hrs.
		thermo-setting resins, based on phenol, urea, furane	1–3		Completion of partially polymerised film by heating or accelerators at room temperature	1–10 min in hot core boxes, 10 min to 1 hr at 150–180°C, 1–8 hrs air setting for small and medium cores
CO_2 gas	as green synthetic sands	sodium silicate solution SiO_2, Na_2O, H_2O (3·3–2–n)	coarse 2·3–3·5 medium 3–4·5 fine 4·5–6		Physical and chemical changes in the silicate films produced by CO_2 or by moisture removal	Gassing for 0·5–1 min or vacuum 5 mm Hg; moulds and cores in most sizes and for most alloys
Shell (C-process)	as green synthetic sands	phenol or urea, formaldehyde resins	cores 3–5 moulds 4–6		As in thermo-setting resins for cores, but using heated metal patterns and core boxes	Patterns at 200–260°C, shell forming 10–20 sec, shell curing 30–60 sec, small to medium moulds and cores

TABLE 8.1 (*Contd.*)

Process	Base aggregate		Binder		Bonding mechanism	Typical methods and applications
	Chemical nature	Particle size range, mm	Chemical nature	Weight % Water %		
Plaster of Paris	plaster moulds $CaSO_4 \frac{1}{2}H_2O$ plaster bonded moulds: as green synthetic sand	fine fine to medium	Plaster of Paris	plaster to water 0·5-1 by weight plaster water solution 20-40% of dry base	Plaster film converted into gypsum by hydration, excess water dried at 200°C	Plaster moulds up to 750°C and plaster bonded moulds up to 1300°C, for small and medium moulds and cores
Precision (investment, lost wax)	as in green synthetic sand; also sillimanite, calcined aluminium silicate, alumina, plaster	precoating 0·05, slurry or stucco, medium to fine	mainly solutions: ethyl silicate $Si(OC_2H_5)H_2O$ silica gels; also sodium silicate, plaster	typical for stucco: ethyl silicate 47%, meths. 41%, H_2O 12%, HCl 0·25%, fine base 150 g	Solid film of silica binder formed by hydrolysis, speed controlled by catalysis and pH, details of numerous variants of solutions in the further reading	Silica binder slowly dried at 100°C and fired at 800°C; choice of base and binder controlled by pouring temperature; mainly small precision moulds; a number of process variants
Cement	as with green synthetic sand	mainly mixed, coarse, medium and fine	mainly calcium, aluminium, silicate: CaO, Al_2O_3; SiO_2	cement 8-10 water 4-6	Hydrolyses to form separate compounds controlled by composition and special additions (cellulose derivatives)	Room temperature setting for 24-28 hrs, fresh mixture for facing; for medium and large moulds and cores

8.6.3 Mould dressings

Mould dressings, when used for surface control of mineral moulds or cores, are prepared on the same principles as the moulding mixtures, except that the mineral base particles are very much smaller than those used for moulds (< 0.001 mm). A widely used base for dressings is carbon (blacking) suspended in water or alcohol with small additions of other binders.

8.6.4 The moulding mixture and methods of mixing

Moulding mixtures that can be obtained by various combinations of minerals and binders have certain characteristic properties which define the essential nature of the moulding process and give the mould, and hence the casting, particular characteristics. However, a feature common to all moulding mixtures is the necessity for mixing the ingredients in such a way as to obtain a mould which is satisfactory from an engineering point of view as well as having optimum metallurgical properties.

One of the common difficulties encountered with a moulding mixture is that of spreading the binder on to the base mineral. With organic binders of suitable viscosity, mixing in a blade mill is sufficient. With clay binders, the spreading of the clay on the mineral base requires both the presence of moisture and the action of a more positive force, such as in roller milling. To prepare a moulding mixture, ingredients, mixing forces, speeds, methods and times of mixing, and bench life of the mixture must all be considered.

8.7 MOULD MAKING

The main metallurgical objective of making up and controlling various moulding mixtures is to enable the moulding process to be carried out satisfactorily as well as producing repeatedly the required properties in the castings. The necessary moulding and mould properties will be more readily appreciated if the elements of mould making are first introduced, since they bear an important relationship to the properties of the castings obtained.

8.7.1 Non-compacting methods

When a binder contains so much liquid that the mineral-binder mixture can flow under the force of gravity, the mould or core is made by pouring the mixture over the pattern or into the core box, e.g. in investment in the lost wax process, or in the pouring of Plaster of Paris or foamed sand-cement mixtures. To achieve proper settling, the mould poured may be vibrated on a vibratory table. An alternative method is to immerse the pattern in the liquid mixture, following by stuccoing with the unbonded mineral, fluidised with a gentle air pressure. By repeating the process a ceramic shell is obtained. A mineral-solid binder mixture can be made fluid ('fluid sand') by adding a suitable foaming agent requiring no ramming on the pattern. In addition, a sand-clay mixture can be made into a slurry with extra water, and such mixtures can be moulded with a rotating profile pattern (strickle) to produce symmetrical shapes (loam moulding).

8.7.2 Compacting methods

With solid binders plasticised by the addition of water, and with viscous organic binders, the application of force is normally required for the moulding mixture to take the shape of a pattern or to fill the core box. With patterns, the following compacting methods are most widely used: ramming, pressing

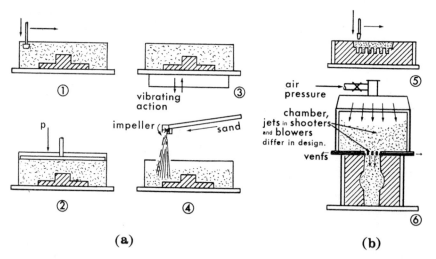

FIGURE 8.8 Compacting methods used singly or combined for (*a*) moulding and (*b*) core making: (1) ramming, (2) pressure, (3) jolting, (4) slinging (throwing), (5) ramming, and (6) blowing and shooting (presssure chambers and jets for shooters and blowers differ in design).

or squeezing, jolting, and throwing or slinging; with core boxes: ramming, pressing, blowing and shooting, Fig. 8.8. Whilst human hands are still one of the most versatile tools for compacting moulding mixtures by ramming on to the pattern or into a core box, hand moulding is mostly retained in foundry practice only where low capital cost rather than a high rate of productivity is the over-riding factor, such as with a single or a few castings off, for complex or heavy designs, or for art castings. A weakness of hand moulding is the lack of reproducibility of moulds of identical properties. Machine compacting has called for a continuous application of engineering ingenuity in designing moulding equipment, varying from simple lever actions to a complex assembly of forces which may be applied by automation. From the point of view of mould making, therefore, the essential property of a moulding mixture used for compacting is its plasticity or ability to take the shape of a pattern. This is controlled partly by the properties of the base (shape, size and size distribution), but mainly by the properties of the binder. The next most important property is that of strength, i.e. the ability of the compacted mould or core to sustain its own weight in the construction of the mould. With sand–clay mixtures this strength can be achieved by compaction (green strength). Some organic binders, particularly often used for cores, have such a low compacted green strength that the core has to be mechanically supported (internally by core wires and externally by core supports) before final hardening (core baking). Hence the advantage of core binders which allow hardening in the core box by thermal or chemical hardening, Table 8.1.

From the point of view of mould or core properties, the different moulding mixture and compacting actions can be examined on a common basis by comparing the mould surface quality, the degree of compaction or ramming density, and the uniformity of ramming. The importance of these characteristics will become clearer in sections 8.8 and 8.9, but in general there is no single moulding machine which could combine the desired optimum mould properties with productivity and casting design variations. The large range of moulding equipment at present in use in foundries is the result of varying needs as well as of continued improvement of moulding operations.

8.7.3 Mould and core assembly and interchangeability of moulding mixtures

Developments of different moulding mixtures and methods of compacting have aimed at improving the properties of the mixtures and shortening the mould making process, so that moulds or cores can be produced and used with as few operations as possible. Moulds made in mineral-clay mixtures can be used without further treatment (green moulds) or after skin drying or more prolonged drying (to increase the strength of moulds for heavier castings) as in dry mould practice. Resin, heat-hardened (shell) and CO_2-hardened moulds and cores can be used immediately after assembly of the moulds and cores. Mixtures based on organic binders may or may not require heating; cement bonding requires standing for a few days, but plaster requires drying out, and high temperature firing is necessary with ethyl silicate. The object of all these variations in the handling of moulding mixtures is to reconcile the conflicting demands of the metallurgical and engineering factors arising in mould filling.

It is clear from the general properties of the moulding mixtures discussed so far that some can be used for both moulds and cores, whilst others are suitable for either one or the other. The main deciding factors are the casting design, the weight of the casting and the composition of the alloy. For example, mineral–clay mixtures are mainly used for moulds, but when partly or fully dried they can sometimes be used for cores, as in bell casting where their removal after casting presents no problems. Organic binders are mainly used for more intricate cores (8.5.2), but they can also be used for moulds. In the latter case, all the various parts of a mould are made in core boxes and, after hardening, they are assembled to form the external part of the cavity of the mould, as well as the internal parts or the cores proper. This is termed the 'all-core' process, often used for casting motor car cylinder blocks. Similar methods can be used with CO_2-hardened, cement and shell moulds and cores. Casting design, properties required and productivity factors can sometimes be reconciled by using moulds made of one or two mixtures only, or several mould mixtures may be necessary for mould or core parts of a single mould. Such interchangeability can be extended to the gravity die casting field, where it is sometimes necessary to use cores made of mineral mould mixtures. This is true also of plaster and ethyl silicate mixtures, though in general these are less frequently used in combination with other processes.

8.7.4 Design of the running and feeding system in a mould (gating system)

In addition to the cavity, the mould must contain channels and openings for running the liquid metal into it and feeding the casting during the process of solidification, as discussed separately in Chapters 9 and 10. The design and construction of the gating system is an integral part of the design of patterns and moulds. The necessary channels or gates and reservoirs or feeders can be cut out manually, but whenever possible the patterns are designed to include the parts which will leave the correct channels of the gating system in the mould at the same time as the moulds are being made. There are many exceptions to this procedure, and frequently for metallurgical or production reasons the parts of the gating system are moulded separately and assembled subsequently with the rest of the mould. In such cases the running or feeding system can be moulded in different mixtures from the rest of the mould; for example, CO_2–sand or even fired refractory channels are used for the running components of the sand moulds for steel castings, and a wide range of materials can be used for feeder walls (Chapter 10).

8.8 MOULD AND CORE BEHAVIOUR IN THE CASTING PROCESS

8.8.1 Mould and core gases

Under normal circumstances the mould cavity is full of air at atmospheric pressure when pouring starts. This air is expelled from the mould cavity during filling by the incoming liquid metal, either through the mould wall itself or through small vents or larger risers. Permeability of the walls is thus an essential feature of most mineral base moulds. In addition, most binders used in moulding mixtures generate gases as a result of being heated by the incoming molten metal; for example, water vapour from the moisture of the clay, and products of combustion and volatilisation of the organic binders. These gases usually escape through the permeable mould and core walls, and partly through channels or vents. Otherwise the mould gases from one of these sources might lead to various types of casting defect, as discussed in Chapter 11. On the other hand, ethyl silicate bonded and subsequently fired moulds have, like most metal moulds, to deal with the air removal problem only.

8.8.2 Strength (compression)

Moulds and cores must be strong enough to support their own weight without deformation (usually a compression force), and the cores may sag if supported horizontally at the ends only. In addition, the incoming molten metal imposes considerable pressure on the mould and core walls. The total metallostatic pressure for a liquid metal of density ρ at a depth h for an area A (in appropriate units) is

$$p = \rho h A$$

It can readily be seen that for deep moulds or cores (h) and large areas (A), the liquid pressure can reach considerable values; hence the necessity for holding mould and core parts mechanically. In small size moulds this is done by locking moulding boxes, and for heavy moulds or cores by using, in addition, different types of mechanical reinforcement. The bonded and compacted mould or core has to meet these strength requirements during the casting process. This explains the reason for drying clay-bonded sand moulds when making larger castings.

8.8.3 Expansion

In addition to the gas and metallostatic pressures, the mould walls also have to withstand the stresses produced by non-uniform thermal expansion of the base mineral, augmented by its phase transformations (8.6.1), the hot mould faces expands relatively to the cooler parts of the mould. A failure of the mould or core surface, in the form of a crack which fills up with liquid metal, gives a poor surface to the casting (scab). A different cause of mould expansion is found with some alloys which tend to expand during solidification, e.g. grey cast irons. Here the metal tends to expand the mould cavity by pushing away the mould walls, resulting in an enlargement of the mould cavity and the forming of a larger casting than would have been obtained with a rigid mould. Strength, rigidity and hot plasticity of moulds and cores are therefore an inter-related group of factors which may affect the metallurgical quality of the cast product.

8.8.4 Collapsibility

Once the solidifying casting has sufficient strength to retain its own shape without external support from the mould, the strength of the mould walls, which was desirable at the pouring stage, becomes undesirable, if too high, from the metallurgical or production engineering point of view. The major reason for this is the contraction of metals during cooling, which if opposed by the mould or core may lead to defects in castings such as warping, cracking and residual stresses (6.8.2). For production reasons, castings have to be freed readily from the surrounding sand, and the core has to be removed from inside the casting. Hence collapsibility rather than retained strength of mould or core materials becomes advantageous. This applies to most production processes, and there is no ideal moulding mixture which could meet the requirements of both strength and collapsibility for all the possible applications. Thus an organic binder is normally preferred for cores, since it has the best compromise properties, whilst the use of clay binders is more general for external sand mould walls.

8.8.5 Metal penetration and mould surface phenomena

An ideal mould surface would be physically smooth and chemically non-reactive with the liquid metal. Ceramic moulds are the nearest to this ideal; the wax pattern is coated with a fine, highly refractory dressing, backed by coarser materials, and the binder gases are eliminated during mould firing. In all the other processes, particularly in sand moulding, for production and metallurgical reasons coarser, less pure mould-making materials are used, and mould dressings are not usually applied. Therefore there is the possibility of metal penetration between the mineral particles of the mould wall, and of the metal reacting with some of the solid or gaseous constituents of the moulding mixture. This can lead to metal contamination as well as to difficulties in cleaning the surfaces of the castings. Metal penetration is governed by the particle size and chemical nature of the mineral base and the binder and by the properties of the liquid metal poured. As the particle size of the mineral base also controls to a large extent the permeability of the mould, it is usually selected to give optimum surface quality with the required permeability. Consequently, for lower melting temperature range alloys, where permeability requirements are not high because of the slow rate of gas evolution, finer mineral base materials can be used. The opposite holds for steel castings. Metal penetration can also be controlled by the application of mould and core dressings or by special additions to the moulding mixture, such as crushed coal (5% to 8%) to the moulding sand in grey iron founding. A carbonaceous addition, in this case, not only reduces the tendency of the metal to wet the sand grains and thus to penetrate into the intergranular space, but also—due to the volatilisation and combustion of the coal products—the pressure in the interspace between sand grains opposes the metallostatic pressure of the metal. The possibility of improving casting surfaces by special additions to the moulding mixtures has to be considered in relation to the moulding and casting process as a whole; for example, coal cannot be added to the moulding mixtures for steel castings, as this would lead to carburisation of the steel at the casting surface. The mould, and hence the casting surface, in any given moulding process is controlled by the basic properties of the moulding mixture, and by special additions or the application of mould and core dressings. In general, mould and core dressings are used to deal with the problems of high searching alloys, liquid metal wetting, and mineral fusion and casting surface smoothness requirements.

8.8.6 Résumé of moulding mixtures

It is evident from the general discussion of moulding mixtures that in a single type of casting process one or more different moulding mixtures can be used. For example, a sand foundry can operate with one moulding sand only, known as a unit sand. After knocking out the castings, this sand is reconditioned, its moulding properties being largely restored by the addition of moisture and in most cases new sand and clay, to balance the gradual accumulation of any 'dead' sand. Alternatively, a foundry can use new, better quality, sand for facing the mould and the old used sand for backing the facing sand, giving a two-sand system. With cores too, one or more mineral base sands can be used in a single foundry and often several types of binder. At the knockout of castings, core sand is normally kept out of the moulding sand, and in general it is more economic to use new core sand than to recover for re-use the mineral base of the core mixture.

Such variations in sand and core mixture practice are more frequent in sand moulding than in more specialised moulding processes, as for example shell moulding, where the selection of the base and the binder leads to few possible process variations, and reconditioning of the used mixture after the knockout of castings can seldom be justified economically. Consequently, the metallurgical problems of process control that have to be considered with various sand mixtures differ from case to case.

8.9 PROCESS CONTROL OF PROPERTIES OF MOULDING MIXTURES

From the point of view of moulding process control, it is convenient to discuss moulding mixtures according to whether they are, or are not, normally restored for subsequent re-use once the solid castings have been removed from the moulds. Mostly the binder is damaged to some extent by the heat of the molten metal, and the problem is whether it is economic, as well as metallurgically feasible, to recover the mineral of the mixture and to add a fresh binder.

The economic factors involved vary, but the average current practice is that sand–clay mixtures are re-used after reconditioning, but other moulding mixtures generally are not. Core sands based on organic binders are as a rule discarded, but some core sand inevitably gets mixed with the moulding sand at the knockout station in a sand foundry. Sodium silicate, CO_2-hardened mixtures, and shell moulding mixtures are not normally re-used, though in principle the binders could be satisfactorily removed to recover the mineral base. Lost wax and cement moulding mixtures are frequently crushed after removal of the casting and graded for subsequent re-use.

8.9.1 General considerations of properties of moulding mixtures

The object of studying the properties of minerals and binders in relation to the metallurgical and engineering principles of metal casting is to provide a scientific basis on which an understanding of the moulding processes, as well as their development, can be built. Metal casting was, however, developed and practised much earlier than the science of materials. Consequently, the general approach to the study of moulding mixtures in the past has been empirical and limited to the development of tests and the study of properties, as the most ready way of dealing with the problems immediately in hand. The most important criteria are permeability, strength, plasticity and high temperature behaviour. As a result a number of empirical tests and methods have been developed which have contributed largely to a better understanding of and various improvements in several casting processes. Scientific studies of minerals, binders and mixtures are at present being made, but such data as available are not used yet as the basis of moulding mixture controls. Current practice (which will be summarised in the subsequent paragraphs) of testing and control of the properties of moulding mixtures, aims at the correlation and analysis of data obtained from empirical tests which have proved themselves by experience. It may well be that in the future some or most of these tests will be replaced by tests based on more scientific concepts of the evaluation of the properties and behaviour of moulding materials and mixtures.

8.9.2 Materials used once only

The properties of moulding or core mixtures are in this case mainly controlled by the properties of the incoming raw materials. The mineral base can be examined for grading (sieve tests), impurities and moisture content, while the binder can be tested for chemical composition, viscosity or density. Moulding mixtures may be tested for compression, tension or bending strength, in the green state or after hardening or baking at room temperatures. They might also be examined for rate and volume of gas evolution at different temperatures. Typical of this group are oil or mineral bonded mould and core mixtures, apart from clay.

8.9.3   Materials partly recovered (mineral base only)

The difference between this case and the preceding one is only marginal. In addition to the new raw materials, a proportion of the complete moulding mixture is made from mineral recovered from the used and crushed moulds (cement and lost wax processes). In all major respects the methods for testing and control of properties are those of new mixtures with a special emphasis on the importance of re-used material.

8.9.4   Materials fully recovered and moulding sand reconditioned (sand-clay mixtures)

The examination and testing of fully recoverable and reconditioned moulding mixtures has received a good deal of technical and some scientific attention, largely due to the importance of mineral-clay mixtures in the sand foundry industry. The scientific background and the empirical nature of present testing methods will therefore be considered at greater length than in the previous two instances.

Permeability. If a given weight of a clean mineral of uniform particle size d is packed into a tube, without applying any force apart from gravity, then the volume taken by the mineral will be V and the bulk density of the sample ρ in Fig. 8.9a. If the sample is compacted by ramming (e.g., a falling weight W) or by applying pressure p, then this volume will decrease and the bulk density increase. If the sample were fully fused to a solid mass, the volume would decrease still further, and the bulk density would increase to a value approaching the theoretical maximum for the mineral. The change in the bulk density of the original uncompacted sample can be used as a measure of the change in the amount of porosity or permeability of the sample. It is clear that the permeability decreases as the bulk density increases, and it is interesting to analyse the effects of important variables of the mixture on the value of the permeability. The most important of these variables are: size, shape and variations in grading of mineral particles; amount, distribution and type of binder and additives; and the magnitude and method of application of the compacting force. If, for example, the experiment previously described were repeated using a finer particle size d_1, then the volume occupied would be V_1, and it would be found that $\rho_1 > \rho$, showing that fine sand would be less permeable for the comparative conditions of testing. If, on the other hand, a given amount of clay and water were added and mixed with these minerals and the experiment of tube packing and compaction repeated, then the following observations would be made. Under the force of gravity alone, the same sample weight would give a greater volume, V_2, than the volume of the unbonded mineral, and $\rho_2 < \rho$, i.e. the bulk density of the sample decreases by bonding. Conversely, if a compacting force, e.g. a falling weight, were used to compress the two samples, made of coarse and fine sand respectively and bonded with the same amount of clay and water, then the bulk density would increase faster in comparison with unbonded sand, as shown in Fig. 8.9a.

The examples discussed indicate that the permeability of a compacted sand in a mould will depend on a number of factors, such as grain size, shape and size of distribution of the mineral base, the kind and amount of clay used, the water content, and the method and extent of compacting, Figs. 8.9b and c. It is also clear that, when dealing with the permeability of moulding mixtures in practice, it is convenient to use a 'comparison standard' to which other

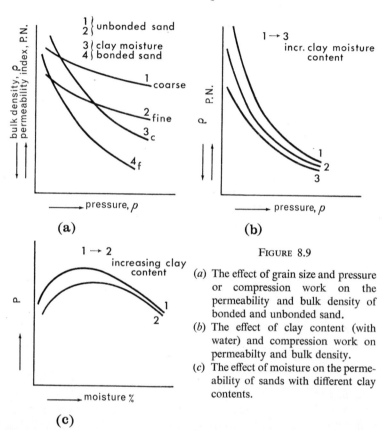

FIGURE 8.9

(a) The effect of grain size and pressure or compression work on the permeability and bulk density of bonded and unbonded sand.

(b) The effect of clay content (with water) and compression work on permeabilty and bulk density.

(c) The effect of moisture on the permeability of sands with different clay contents.

permeability measurements can be related. The generally adopted unit is known as the permeability number or index obtained on a sample prepared and tested in a standardised manner (American Foundrymen's Society or A.F.S. sand testing procedure, see text for further reading).

Compression strength. In general, moulding mixtures are subjected to compressive forces. If the experiments on coarse and fine sands described in the previous paragraph were repeated and the samples subsequently tested for compression strength, the results obtained would resemble those shown in

Fig. 8.10*a*. The development of a measurable strength with unbonded sands requires such high pressures that it is not generally a practicable method of moulding. With the addition of some clay and water, the compression strength of a sample can be varied, as shown in Fig. 8.10*b* and *c*. The same variables which affect the permeability clearly show an effect on the compression strength.

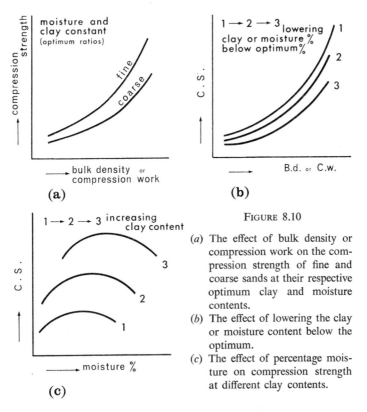

(a)

(b)

(c)

FIGURE 8.10

(*a*) The effect of bulk density or compression work on the compression strength of fine and coarse sands at their respective optimum clay and moisture contents.

(*b*) The effect of lowering the clay or moisture content below the optimum.

(*c*) The effect of percentage moisture on compression strength at different clay contents.

Principles of current sand testing practice and controls. The objectives of sand testing practice are considered in 8.9.5. The guiding principle in procedure has been that of testing a representative sample of prepared moulding mixture for those properties that have been found in practice to be essential for production control in a given foundry process. Tests are carried out on the behaviour of sand prior to and during moulding, and the results of such tests are related to the properties of the castings produced. The present internationally used procedure for sand–clay mixtures includes tests for 'free' moisture content, compression strength and permeability. Other properties tested less frequently are mould hardness, plasticity or mouldability, effects of heating on sand strength and shatter behaviour.

The free moisture content is determined either by weighing the loss in moisture at 110–120°C of a 5–10 g loose sand sample, or by using a direct moisture reading instrument. The significance of the moisture content of a sand–clay mixture is apparent from Figs. 8.9c and 8.10c.

Both the compression strength and the permeability are determined from a compacted sample prepared in a 2 in. dia. tube by dropping three successive times, under gravity, a 14 lb weight from a 2 in. height, and selecting the initial weight of the loose sand so that the height of the compacted sample $H = 2 \pm 1/32$ in. The permeability number, P.N., is obtained by measuring the time t required for a given volume of air V cm^3 to pass through the cross-sectional area A of the sample at a given pressure p in cm of water:

$$P.N. = \frac{V \times H}{A \times p \times t}$$

The permeability apparatus is usually designed for reading directly the value of P.N., i.e. the rate of air flow, under the standardised measuring conditions. The green or dry compression strengths are obtained by direct measurement of the compression break load. The above procedure can also be applied to core sand mixtures, but the tensile and bend tests of specially designed test pieces are used more frequently, being more closely related to core behaviour in moulds.

Other properties. Mould hardness, plasticity (mouldability), the effects of heating on strength, shatter testing and several other tests sometimes used in moulding and core sand testing and control are described in the literature recommended for further reading.

8.9.5 Significance and application of sand testing data

Experience has shown that without sand testing it is difficult or impossible to control the production of moulds and castings with consistent properties. While this remains the major reason for sand testing in foundries, testing practice has also developed in other directions.

Incoming raw materials such as sands or binders are tested to control the desired acceptance standards. The tests used are essentially for the identification of the chemical or physical properties of materials. After mixing or milling of a new or reconditioned moulding mixture the usual tests are for percentage of moisture, permeability and compression strength (A.F.S. procedure), but some additional tests may also be used. The tests are carried out at regular intervals during production, and the results graphed and often statistically analysed (Chapter 11). The basic assumption of such testing practices is that the properties measured are related to and can be used to control the behaviour of sand in moulding and in casting. Consequently, once satisfactory properties of a moulding mixture have been established, tests are only carried out as a process control. In addition to this routine testing various sand tests are used in foundries for the development of new moulding mixtures. Current sand testing procedure covers most of the problems encountered in practice consistent with presently accepted quality standards for cast metals. The demand for a more scientific approach to sand testing may follow an increasing demand for more rigid engineering evaluation of the properties of cast products. It is important to note that sand properties are evaluated for the mixture before moulding, but the properties of compacted sands in moulds are seldom measured and could not be predicted from the routine test data obtained on the moulding mixture after milling. Furthermore, the arbitrary design and preparation of the standard test piece are unsuitable for quantitative evaluation of really fundamental properties of moulding mixtures.

The economic and production significance of sand mixtures in foundries can readily be seen from the fact that the quantity of sand used may be in the ratio of up to 10 or 20 to 1 by weight of the castings produced. A sand foundry can be looked upon as an interplay of two systems—the metal system and the sand system—each proceeding along a definite flow line, but each involving engineering, production and metallurgical considerations.

8.10 METALLURGICAL SIGNIFICANCE OF MOULDS AND MOULDING MIXTURES

The primary function of a mould is that of obtaining castings of required shape and dimensions. On the other hand, the quality of castings is largely controlled by metallurgical phenomena occurring at the surface and inside a casting. Some of these phenomena are strongly affected by the moulds and their properties.

8.10.1 Surface phenomena

The ideal surface of a casting is perfectly smooth and free from any blemishes. In practice, surfaces of castings display various degrees of roughness, due to the surface condition of the mould, and temperatures, flow and chemical reaction phenomena at the metal-mould interface. Surface roughness is a detailed reproduction of mould surface roughness, dependent on the surface tension of the metal and pressure conditions in the mould during filling. Further, the solidification at the mould surface, combined with turbulence during the filling of the mould, can give rise to specific types of surface roughness, such as rippling and cold shuts. Surface roughness is therefore primarily controlled by the choice of mould materials, the condition of the mould surface, mould dressings if used, and by the mould and pouring temperatures. In addition, turbulent flow during filling can lead to a number of surface imperfections: surface oxidation and drossing (particles of metal embedded in an oxide film), detachment of mould surface material and its occlusion at the surface or in the casting, and occlusion of mould gases (Chapter 9). Chemical reactions at the metal-mould interface occur whenever the reaction energy and kinetic factors (Chapter 3) are favourable. For example, the rate of reaction of magnesium with oxygen or water vapour in the mould is so fast that special inhibitors, such as boric acid (2%–4%) and sulphur (0·5%–1%), have to be added to the moulding or core sand mixtures or to mould dressings in the casting of magnesium alloys. The inhibitors consume some of the available oxygen and provide a protective film at the surface of the alloy. Other partial protection from oxidation can be obtained with a reducing atmosphere in the mould provided by combustion of volatile mould dressings or ingredients in the mould mixture, as in ingot casting. Oxygen and water vapour from the mould atmosphere can lead to surface oxidation and gradual solution of oxygen or hydrogen in the alloy (Chapter 4). Another type of reaction at the mould surface is that between mould material, such as sand, with one of the metal or non-metallic constituents of the alloy. For example, the reaction between iron oxide on the molten iron and the silica of the moulding sand is the main cause of sand adhering to the casting surfaces (sand burning). This, combined with metal penetration, largely controls the surface quality of steel castings.

With grey iron and steel moulds (dies) or 'chills' inserted into sand moulds (Chapter 10) when the mould face temperature exceeds 700°C, carbon from the alloy of the mould reacts with the oxygen, if present at the interface, to give CO or CO_2 gas which blows into the skin of the casting. Such a surface defect can be prevented either by mould temperature control or by chemical methods, one of these being the use of a mould dressing containing fine aluminium powder to 'tie up' the oxygen (Chapter 3).

Thus, the metallurgical control of casting surfaces involves mainly the control of some physical and chemical phenomena at the mould interface. Whilst the metallurgical diagnosis of a particular surface defect is not as a rule difficult, the problem of its prevention has to be considered in conjunction with the costs of cleaning the casting surfaces compared with the costs of more elaborate methods of controlling the surface quality in production.

8.10.2   Rate and direction of cooling

The rate of cooling during solidification of a casting affects the type of cast structure obtained and the development of the solidification fronts and zones (6.7.2). The direction of solidification fronts affects the formation of shrinkage cavities during the solidification and feeding of castings. Cavities in castings are normally classified as defects, and because of the great importance attached to their elimination they are considered separately in Chapter 10. The importance of the structure of micro- or macro-grains, on the other hand, varies from one type of casting or alloy to another (Chapter 7). The way in which this problem is affected by the rate of cooling in moulds is discussed in the next paragraph.

Theoretical and applied evaluation of the cooling rate. The most useful definition of the cooling rate is identical with that of the solidification or growth rate, i.e. the rate of movement of the solidus isothermal away from the mould wall. This can be evaluated and verified for specific experimental conditions (6.2 and Fig. 6.2a). For castings in general, heat transfer calculations are much too complex under the cooling conditions and with the casting designs which prevail in practice. Castings cool under the combined effects of several types of heat transfer mechanism—conduction, convection and radiation—so that the solution of a cooling problem is mathematically cumbersome and difficult to define precisely. Heat transfer calculations have, however, been carried out for simple shapes and for idealised conditions, such as semi-infinite plates (10.3.2), infinite cylinders or spheres. With some simplifications of boundary conditions and heat transfer methods, such solutions can be applied to simple castings, e.g. for the analysis of cooling continuously cast ingots. However, most of the progress made in cooling rate evaluation has so far been in the development of various experimental techniques rather than in rigorous application of heat transfer calculations to casting and mould designs.

Cooling rate measurements. The simplest and oldest method of evaluation is to measure the actual thickness of the casting skin by decanting residual liquid (slush casting). This method is experimentally rough and clearly limited to pure metals only (6.9). A more powerful, though rather expensive, method is to insert thermocouples at different points in the casting in the direction of solidification and record the temperature changes. For castings made in mineral moulds an equally useful method is to place the thermocouples in the mould itself. The results obtained by either method have been used to compare the cooling rates of metals cast in different moulding materials (chilling power) and to verify some of the idealised treatments of heat transfer, 10.3.2. Finally, direct heat transfer measurements can, in some cases, be replaced by analogous but experimentally simpler techniques. The electrical analogue method, for example, can be used to examine in greater detail numerical effects of heat transfer variables, once the boundary conditions have been experimentally established by one of the direct methods.

Application to ingots. Owing to their simple geometry, ingots could be considered an obvious choice for theoretical and applied heat transfer studies. The cooling methods used for ingots are illustrated in Fig. 8.11. With solid moulds, the heat is transferred largely by conduction, but in this instance a purely analytical solution is difficult owing to the corner and end effects and the uncertainty of the interface temperature. Consequently, most heat transfer data for solid ingot moulds have been obtained from direct cooling curve measurements, supported by analogue techniques. Water-cooled moulds are simpler in that the outer mould wall temperature is maintained constant, but in most practical cases the moulds are filled slowly and therefore the metal mould interface temperature is even more difficult to define than that of solid moulds. The simplest form of heat transfer is encountered in continuous casting, because the boundary conditions can be clearly specified. For idealised heat transfer conditions, as shown in Fig. 8.11, it is clear that the rate of heat transfer in solid moulds is mainly controlled by the chilling power of the mould, and by interface conditions; with water-cooled moulds the controlling parameter becomes the conductivity of the

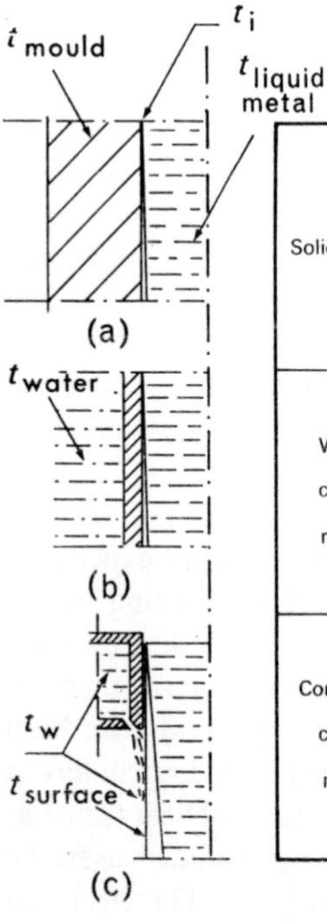

		Idealised conditions	Actual conditions
Solid mould	Temperatures	t_i constant t_L and t_m assumed of known behaviour	all temperatures variable, t_L mainly across
	Heat transfer	conduction	air gap forms at the interface; convection and radiation arise
	Geometry	semi-infinite conditions, instantaneous filling	corner and end effects
Water-cooled mould	Temperatures	t_i and t_w constant, t_L assumed of known behaviour	all temperatures variable, t_L across and vertically
	Heat transfer	conduction	air gap and mould dressings; conduction and convection
	Geometry	semi-infinite conditions, instantaneous filling	corner and end effects, slow rate of mould filling
Continuous casting mould	Temperatures	mould as above; below mould t_s assumed low	variations more predictable than above
	Heat transfer	mould as above; below mould conduction	below mould approx. steady conditions
	Geometry	semi-infinite conditions	bulk of heat transfer below mould; approx. semi-infinite conditions

FIGURE 8.11 Cooling of ingots in (a) solid, (b) water-cooled, and (c) continuous moulds

mould wall, and for continuous casting the conductivity of the alloy cast. The corresponding increase in the cooling rate is accompanied by changes in the cast structure (6.2.1 and Fig. 6.2).

Application to shaped castings. With metal moulds, the transient nature of heat transfer, combined with the complex factors of casting design, makes any attempt at general heat transfer analysis very difficult; the direct recording of cooling curves is the most practicable approach. Mineral moulds, on the other hand, have such a slow cooling rate that, apart from casting design complications, heat transfer analysis becomes more feasible, 10.3.2.

Significance of the cooling rate. With ingots, the main structural features, the grains, cavities and inclusions, control the ingot plasticity during subsequent working. With some alloys, e.g. rimming steels, none of these structural parameters seriously impairs the hot working plasticity of the alloy. Consequently, with alloys of this type the importance of the rate of cooling is on a par with production or economic factors, and the metallurgical quality of the finished product is controlled by the melting and fabrication processes. With some other alloys, e.g. high strength aluminium base alloys (Duralumin types) the required alloy plasticity can only readily be obtained with the continuous casting process, which achieves the required cast structure. For many types of ingot, cavities can be closed up during deformation, and the structure of the alloy constituents is then of greater importance.

The cast structure in shaped castings, unless the alloy can be heat-treated, determines permanently the structure-sensitive properties on which the use of castings often depends and the control of the structural features is all the more important. As an increase in cooling rate usually allows a better control of all the structural factors, metal mould castings are likely to have better metallurgical quality than sand castings. Mineral mould castings, owing to their intrinsically slow cooling rate, are the most difficult of all metallurgical products to produce consistently, to ensure that the required grain structure and freedom from cavities and inclusions are obtained.

8.11 SUMMARY

By definition, castings are products made in moulds. As there are a number of possible moulding processes, which can be operated singly or in combination, an essential primary problem of founding is that of selecting the optimum moulding process. This is done on the basis of a consideration of casting design, properties of castings and production requirements. The second problem, which is equally frequent, is that of operating a given moulding process in such a way as to achieve optimum metallurgical and production results. In this respect casting processes are akin to other processes of manufacture.

Mineral mould making is at one and the same time the pride and the Achilles' heel of founding, the former because mould design and construction often require a superb grade of human skill, the latter because the mould is the root cause of the dust, fumes and general mess often seen in foundries. However, these conditions are evidence of human nature rather than an inherent characteristic of foundry processes as such.

FURTHER READING

VAN VLACK, L. H., *Elements of Material Science*, Addison-Wesley, Reading, Mass., 1959.

DAVIES, W., *Foundry Sand Control*, The United Steel Co. (G.Br.), 1950.

Foundry Sand Handbook, American Foundrymen's Society, 1953.

DIETERT, H. W., *Foundry Core Practice*, American Foundrymen's Society, 1950.

BRIGGS, C. W., *Fundamentals of Steel Foundry Sands*, Steel Founders Society, Cleveland, 1959.

Patternmakers Manual, American Foundrymen's Society, 1953.

PETRZELA, L. and GAJDUSEK, J., The CO_2 process, *Trans. Amer. Found. Soc.*, 1962, **70**, 65.

DOEHLER, H. H., *Die Casting*, McGraw-Hill, New York and Maidenhead, 1951.

BARTON, H. K., *The Die Casting Process*, Odhams Press, London, 1956.

CADY, E. L., *Precision Investment Casting*, Reinhold, New York, 1948.

FUCHS, W., Die Anwendung des Zementsandes, *Giesserei*, 1954, **41**(11), 278.

MOUNTAIN, K. L., Intricate plaster mould castings, *Foundry*, 1955, **83**(11), 100.

STOCK, C. M. and BOWNES, F. F., Aspects of sand compaction by moulding machines, *Br. Foundryman*, 1961, **54**, 428.

IRMANN, R., *La Fonderie d'Aluminium*, Dunod, Paris, 1957.

NEWELL, W. C., *The Casting of Steel*, Pergamon, Oxford, 1955.

SCHWARZMAIER, W., *Stranggiessen*, Berliner Union, 1957.

AITCHISON, L. and KONDIC, V., *The Casting of Non-ferrous Ingots*, Macdonald & Evans, London.

CHAPTER 9

FLOW OF METALS INTO MOULDS

9.1 INTRODUCTION

In the majority of casting processes liquid metal is supplied to a mould from a melting furnace, ladle or crucible, and the nature of the casting process determines the method of pouring and mould filling. The system of channels which introduces the liquid metal into the mould is called a gating system. The design of the system depends on the type of casting process and the casting design. Most systems can be divided into two parts: (*a*) the gating or running components, and (*b*) risering and feeding components, Fig. 8.4*e*. However, the term gating system is often used to imply the running system only, and with some processes this serves the function of the feeding system as well, Fig. 8.2.

9.2 GATING OR RUNNING SYSTEM

The gating or running system is a system of channels along which the metal runs into the mould cavity. Certain types of mould, used for example in ingot or some centrifugal casting processes, can be filled directly from the container supplying the liquid metal, so that they have no true gating system, Figs. 8.1 and 8.3. This method of mould filling, however, may not give an ingot or casting of satisfactory metallurgical quality, and it may be necessary to pour the metal into the mould at a controlled rate through an aperture This is an example of the simplest form of gating system (Fig. 8.3*b*), which is described as tundish pouring in ingot casting. Some shaped castings too can be poured directly through a simple channel, called a downsprue or down-runner, into the mould cavity (Fig. 8.2). In order to obtain a satisfactory metallurgical and engineering quality, with most shaped castings poured into mineral moulds the gating system should have several channels (Fig. 8.4). Apart from metallurgical considerations, the moulding, casting design and production requirements bring other factors to bear on the problem of the design of the gating system. Metallurgically, a gating system should maintain the metal 'quality' during flow, avoiding oxidation of the metal stream and entrainment of gases, slag, flux or mould material. Moulding practice and

8

casting design require a system of channels that simplifies the moulding process, allows the metal to enter the mould cavity at the correct place and speed, ensures the complete filling of the mould, and gives optimum temperature distribution in the casting from the point of view of feeding and of contraction stresses. Production requirements demand economy in the design of the gating system (casting yield) and its easy removal from the finished casting. These are the main reasons for the development and use of different gating systems for different alloys, casting designs and processes, and for different moulding and production methods.

9.3 RISERING AND FEEDING SYSTEM

In a casting process, all the required details of the mould cavity should be filled with the liquid metal. Normally, all gases, including atmospheric air and in some cases gases generated in the mould, must be expelled. This is achieved by the provision of gas escape passages through permeable mould walls and mould vents. But sometimes special channels or reservoirs attached to the castings are used, which not only allow the mould gases to escape, but simultaneously provide extra metallostatic pressure, forcing the liquid to reproduce the mould details more fully (fluidity, 3.4.2). These reservoirs are termed risers. In addition, such reservoirs are even more often required to provide extra liquid to compensate for the volumetric contraction (shrinkage) of the metal in the mould in the liquid state and during freezing (6.7.2). In such instances these reservoirs are termed feeders or headers. As most casting alloys contract while freezing, the application of feeders is more general than that of simple risers. Owing to the general importance of feeding in many foundry processes, this subject is dealt with more fully in Chapter 10.

9.4 GATING ELEMENTS

Numerous types of gating system are used in the various casting processes, and even in a single casting process, such as sand casting, many gating designs can be used. Gating systems used in sand founding are generally more complex and varied than in other casting processes because of their important function in controlling the quality of the castings made. As other gating systems can be readily understood in terms of sand casting systems, this chapter will deal mainly with problems relating to these, and the conclusions reached will then be applied to gating systems used in some other processes.

A gating system used for sand casting comprises certain basic elements, shown in Fig. 9.1. The pouring basin or cup receives the liquid metal and acts as a reservoir from which the liquid metal flows into the remainder of the

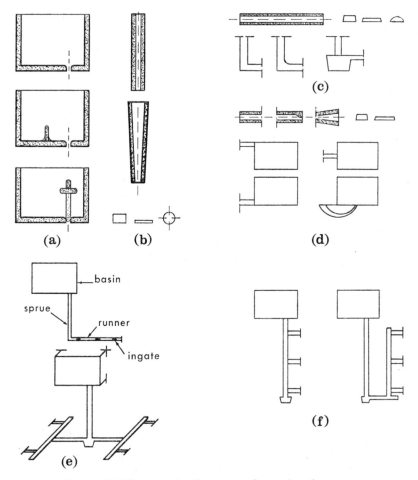

FIGURE 9.1 Elements of gating systems for sand castings

(a) Pouring basins with design features for controlling the start of pouring (cleanliness, vortex control and filling the sprue).

(b) Parallel wall and tapered sprues and their sections (rectangular, slot and round).

(c) Runner bar, its cross-sections and methods of joining to the sprue (sharp or rounded elbows and the sprue-well).

(d) Ingates (parallel, choked, flared) and methods of joining to the casting (top, side—mould parting plane, bottom and bottom horn).

(e) Simple gating system (basin, sprue, bar and three ingates), and a more complex manifold horizontal system.

(f) Vertical step systems.

gating system. It also keeps the liquid metal clean by preventing dross or slag entering the gating system proper. The sprue or downrunner takes the metal vertically from the pouring basin to the next elements, which are usually horizontal, or it may lead directly into the mould cavity, in which case the complete gating system comprises only two elements. Joined to the base of the sprue are the liquid metal distribution elements of the gating system, namely the runner bar and the ingates. From the base of the sprue one or more runners can be directly connected to the mould cavity, the gating system then comprising three elements, basin, sprue and runner, with the runner performing the function of the ingate. However, the number of

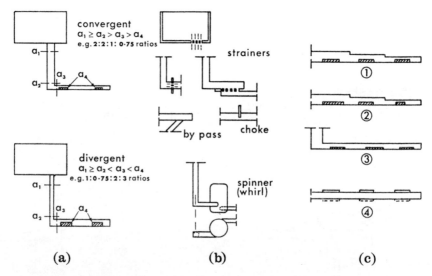

FIGURE 9.2 Control of flow by design of gating systems

(a) Velocity: convergent (choked) high velocity flow, and divergent reduced velocity flow.

(b) Cleanliness ensured by strainers, direction changes, chokes, or by spinning (centrifuging).

(c) Distribution: (1) reduction of runner bar area, (2) combined runner bar and ingate area reduction, (3) gating ratio 1:2:1, (4) positioning of ingates at top or bottom.

pouring basins, sprues and ingates may vary in a given gating system for a single casting of different design, complexity or weight.

Metal flow in the elements of a gating system is dependent on their design. The cross-section of a gating element usually varies from half oval to various rectangular shapes, and its area can be constant, or decreasing (converging) or increasing (diverging) in the direction of flow. Special features are sometimes introduced to achieve specific flow control, Fig. 9.2; e.g., controlling the dross, slag or other impurities being formed or carried into the mould cavity, or reducing the speed of flow, or achieving a proper distribution of flow in the various gating elements.

Gating systems have been developed over many years of foundry practice to achieve the best quality castings with as simple a system as possible. A complete numerical solution of the gating problem for a given casting process is not as yet feasible, due to the lack of fundamental data. On the other hand, a partial understanding of the calculations necessary for improving gating systems has become possible through the application of various hydraulic, metallurgical, casting design and moulding principles.

9.5 PRESSURE SYSTEMS

The flow in gating systems can be induced by gravity, external pressure or centrifugal forces (Figs. 8.2 and 8.3). Gravity systems are used for ingots, most mineral mould casting processes, and for gravity or permanent mould die casting. Differences in the mould, the casting design, and the liquid properties of the alloys used, explain the origin of the large variety of gating systems; they are simplest for ingots, slightly more varied for gravity die castings, and still more complex for sand castings. Low pressure systems ($< 100 \, lb/in^2$) are used with permanent moulds mounted above the level of the liquid metal in a holding furnace, but the gating systems are basically similar to those for normal gravity processes. Higher pressures are used in pressure die casting, where the gating system is more critical and strongly influences the success or failure of the casting process. Simple centrifugal castings, such as discs, pipes and cylinders, do not need any special gating systems apart from those that control the flow of liquid metal from the holding crucible into the mould. Shaped centrifugal castings, on the other hand, require gating systems which control the speed, the condition of the metal and its distribution. This chapter deals only with gravity horizontal distribution systems. Other systems can be understood in principle by studying such gravity systems, but their more detailed treatment is dealt with in the recommended further reading.

9.6 THE HYDRAULICS OF GATING SYSTEMS

Systems of flow channels used in founding are not normally encountered in other branches of applied hydraulics, and only some aspects of theoretical treatment or practical data on the more complex gating systems are given in standard hydraulics textbooks. The hydraulics of simple systems consisting of two or three standard elements are readily amenable to theoretical treatment, but considerable difficulties arise with more complex gating systems, or manifolds.

An essential part of applying the study of hydraulics to gating systems is the calculation of metal velocity and pressure, and of energy losses in the system, using available data on the properties of liquids for the dimensions and shapes of the channels. Energy changes are worked out for a system behaving ideally, and the departure from ideal flow is then determined experimentally. The ideal flow equations are then amended to give empirical flow relationships for dealing with problems encountered in gating systems. The measurement of the degree of departure from ideal flow and of energy losses has received far less attention than the study of idealised flow conditions. Advances made in this direction for some gravity flow gating systems are considered below, but in general the dimensions of gating systems in foundry practice are still determined empirically.

9.6.1   The problems of gating system hydraulics

Hydraulic equations of gating deal with three basic quantities of flow—its velocity v, the pressure p at any given point in the flow system, and the distribution of the flow in a multiple element gating system. The flow velocity is required in order to calculate the mould filling time. Foundry experience has shown that there is an optimum pouring rate in certain casting processes, casting designs and weights, and it is frequently necessary to calculate the rates of flow in advance, or to work out the dimensions of gating systems capable of giving the required flow rates. Knowledge of the pressure distribution in a gating system is useful for two reasons. Firstly, if the pressure at a given section of the gating system is below atmospheric, then the danger exists of air or mould gas being drawn into the liquid stream, which could be detrimental to metal quality. Secondly, for some alloys it is essential that the pressure and velocity in the ingates should be as low as possible, in order to reduce turbulence and agitation, which could cause deterioration in the metal quality. The usefulness of data on the distribution of metal in multiple gating systems is related to the solidification of the casting. Depending on the design of the casting and the casting process, it is frequently desirable that a casting should solidify in selected directions (Chapter 10). To achieve this the metal must enter the mould at the required sections and at the required rates—a problem directly related to the knowledge of distribution of flow in multiple ingate systems.

9.6.2   The hydraulics of gravity systems for sand castings. Types of system

Gating systems used for sand castings can be divided into horizontal systems in which the distribution elements are in a horizontal plane, Fig. 9.1e, and vertical systems where distribution is in a vertical plane, Fig. 9.1f. Horizontal systems only will be considered, as they are used much more frequently and have been studied to a far greater extent. They normally contain one or more of the following gating elements: pouring basin, sprue, runner bar and ingate. The flow hydraulics of each element will be considered separately, but gradually building up a complete gating system.

9.6.3 The pouring basin

Hydraulically, the pouring basin can be looked upon as a reservoir with an exit hole in the base. The liquid metal is poured into the basin from an outside container, and the level is maintained constant by continuous pouring until the mould is full. Several other features in the design of the pouring

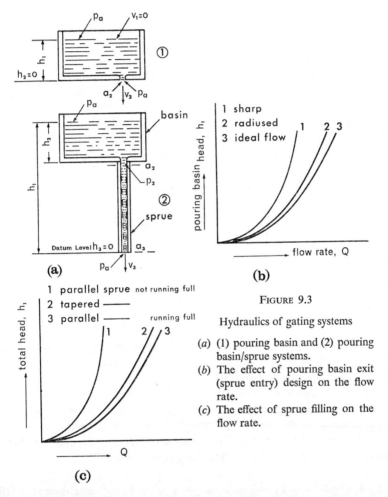

(a)

(b)

(c)

FIGURE 9.3

Hydraulics of gating systems

(*a*) (1) pouring basin and (2) pouring basin/sprue systems.

(*b*) The effect of pouring basin exit (sprue entry) design on the flow rate.

(*c*) The effect of sprue filling on the flow rate.

basin, Fig. 9.1*a*, may arise from the demands of the casting process, and these are discussed subsequently. The two important hydraulic factors are the height of the liquid metal in the basin and the dimensions and design of the exit hole, Fig. 9.3.

A steady state of flow established in the pouring basin ensures that the amount of liquid supplied from the outside keeps the level in the basin constant. Under such conditions the main quantities defining the flow can be related by applying the laws of hydraulics. The first law (constancy of flow volume) states that in the steady state, for incompressible fluids, the volume of liquid passing through any cross-section of the system is constant. This law applied to the pouring basin reads:

$$a_1 v_1 = a_2 v_2 = a_3 v_3 = \text{constant} \tag{9.1}$$

where a_1, a_2 and a_3 are the pouring basin, basin exit and sprue exit cross-sectional areas respectively, and v_1, v_2 and v_3 are the corresponding velocities of flow.

The second law expresses the conservation of energy in the flow system. Considering the flow system of the pouring basin only, several kinds of energy are involved. Potential energy measured relative to the position of the flow exit from the pouring basin is equal to hw, where the head of the metal is h and its weight is w, Fig. 9.3a. The exit level is frequently selected as a datum or zero level for evaluating the potential energy in flow systems. In the flow system under consideration the potential energy is the driving energy of the system which is converted into three other forms:

(a) Kinetic energy $mv^2/2$, which for unit weight of liquid becomes $v^2/2g$, where v is the average velocity of flow through a cross-section of the flow system and g is the gravity acceleration.

(b) The pressure energy pv, where p is the pressure and v is the volume, which for unit weight of liquid becomes p/ρ, where ρ is the density of the liquid.

(c) Loss energy, i.e. energy lost by the system during the flow. The loss energy is usually expressed in terms of a loss coefficient k, and related arbitrarily to the kinetic energy for that part of the system, $k \cdot v^2/2g$. The value of k is determined experimentally for various parts of the system and for varying conditions of flow, in order to obtain a measure of the departure of the flow from the ideal condition.

The law of conservation of energy, applied to flow systems, states that for a steady state of flow the sum of all energies at a given cross-section of the system is equal to the sum of all energies at any other cross-section of the flow. In terms of the energies previously considered, (per unit weight) this law becomes, for a system in which the elements are full,

$$h_1 + \frac{v_1^2}{2g} + \frac{p_1}{\rho} + k_1 \cdot \frac{v_1^2}{2g} = h_2 + \frac{v_2^2}{2g} + \frac{p_2}{\rho} + k_2 \cdot \frac{v_2^2}{2g} = \text{constant} \tag{9.2}$$

This relationship is known as the Bernoulli equation. The actual significance of the h, v and p terms, and the number and values of terms k are most readily explained by considering examples of flow in different elements of the gating systems. For numerical calculations consistent and appropriate physical units for all the quantities must be used.

The Bernoulli equation will first be applied to the flow system of the pouring basin only. Suffix 1 is used in this example for all the energies at the entry, or top level, of the pouring basin, and suffix 2 for the exit level. At the entry the unit weight of liquid has a potential energy h relative to the exit. The velocity of flow at level 1 is zero and the pressure is atmospheric, p_a, but at the exit the velocity is v_2, while the pressure is still atmospheric and the relative potential energy is zero, as level 2 is the datum line in this instance. Between level 2 and level 1 there is one major energy loss, at the liquid entry into the exit hole in the basin. This loss is due to change of direction of flow, which is equal to $k \cdot v_2^2/2g$. The Bernoulli equation, equating all the energies at levels 1 and 2, becomes:

$$h_1 + 0 + \frac{p_a}{\rho} = 0 + \frac{v_2^2}{2g} + \frac{p_a}{\rho} + k_2 \frac{v_2^2}{2g}$$

The velocity of flow at the exit from the pouring basin, v_2, is therefore:

$$v_2 = \sqrt{\frac{2gh_1}{1 + k_2}} = c_2 \sqrt{2gh_1} \qquad (9.3)$$

The value of $c_2 = \sqrt{\dfrac{1}{1 + k_2}}$, which is smaller than unity, is known as the coefficient of discharge. The numerical value of k, which in this case is dependent on the exit hole design, has to be determined experimentally, Fig. 9.3b.

The volume of liquid flowing from the exit of the pouring basin is obtained from the equation of constancy of volume, $Q = a_2 v_2$, where a_2 is the exit cross-sectional area of the liquid stream.

9.6.4 Pouring basin and the sprue system

In the pouring basin and sprue system, the exit from the pouring basin extends into a vertical channel, called a sprue or downrunner. The two basic laws are applied to this system, as with the previous example. While level 1 and level 2 remain as before, the datum or zero level moves to the exit from the sprue (level 3), Fig. 9.3, and we have from the Bernoulli relation for levels 1 and 3:

$$h_1 + 0 + \frac{p_a}{\rho} = 0 + \frac{v_3^2}{2g} + \frac{p_a}{\rho} + k_2 \cdot \frac{v_2^2}{2g} + k_3 \cdot \frac{v_3^2}{2g}$$

where $k_2 \cdot v_2^2/2g$ is the energy loss for the pouring basin entry, and $k_3 \cdot v_3^2/2g$ is the energy loss in the sprue. This latter loss is due to the wall friction and can also be expressed as:

$$k_f \cdot \frac{v_3^2}{2g} \cdot \frac{l}{D}$$

where k_f is the wall friction loss, l the length of the sprue and D the sprue diameter for a round sprue. This requires that $k_3 = k_f(l/D)$. When the system is running full $a_2v_2 = a_3v_3$, so that the Bernoulli equation can be expressed in terms of:

$$\frac{v_3^2}{2g}\left[1 + k_2\left(\frac{a_3}{a_2}\right)^2 + k_3\right] = h_1$$

or
$$v_3 = c_3\sqrt{2gh_1} \qquad (9.4)$$

where $c_3 = \sqrt{\dfrac{1}{1 + k_2(a_3/a_2)^2 + k_3}}$ and has the same interpretation as in the flow equation obtained for the pouring basin only.

Two typical flow patterns can arise in this flow system. Either the liquid entering the sprue contracts and flows out of the sprue without further touching the sprue walls, or the liquid wets the sprue and the sprue runs full. It is clear that the above equation applies only for the latter condition, whilst in the former case the flow system behaves as with the pouring basin only. In other words, the full head h_1 between section 1 and 3 is utilised in the fully running system, but only the head in the pouring basin, h_2, is used when the liquid contracts away from the sprue wall. Consequently the fully running system has a much greater flow rate, as $Q_3 = a_3v_3 = a_3c_3\sqrt{2gh_1}$, whilst in the non-wetting system

$$Q_3 = Q_2 = a_3v_3 = a_2v_2 = a_2c_2\sqrt{2gh_2}$$

The flow efficiency of a parallel wall sprue thus depends on the liquid-to-sprue wall surface phenomena. These are controlled by the nature of the liquid, design and material of the sprue and sprue entry, and—for more complex gating systems—by the design of the gating system following the sprue. In order to avoid inefficient use of sprues in gating systems, they can be tapered, choosing for the exit area a_3 that cross-section which the free falling liquid from exit 2 would take at level 3 in the circumstances of a non-wetting parallel sprue. The condition for a critical taper is therefore that sections 2 and 3 are at atmospheric pressure. The velocities of flow v_2 and v_3 have already been obtained for this flow condition (equations 9.3 and 9.4).

From the continuity equation $a_3v_3 = a_2v_2$, hence $a_3 = a_2(v_2/v_3)$, and by substitution from 9.3 and 9.4

$$a_3 = a_2 \cdot \frac{c_2}{c_3}\sqrt{\frac{h_2}{h_1}}$$

As a first approximation, ideal flow without energy losses can be assumed, in which case $c_2 = c_3 = 1$,

and
$$a_3 = a_2\sqrt{\frac{h_2}{h_1}} \qquad . \quad (9.5)$$

It follows from these considerations that a parallel sprue which runs full and a critically tapered sprue, which runs full by definition of its design, will have a similar flow efficiency for the same exit area at section 3, because in both cases the flow is exit-controlled at section 3. Small differences may, however, arise through the differences in the energy losses in the two cases. On the other hand, a parallel sprue not running full will have a much smaller flow efficiency than a tapered sprue of the same exit area, $a_3 = a_2$, because in this case the flow rate in the parallel sprue is controlled at section 2, Fig. 9.3c.

In addition to comparing the flow rates obtainable from parallel and tapered sprues, it is also of interest to examine the pressure values at section 2 for both cases. The design of the critically tapered sprue has been calculated for atmospheric pressure at sections 2 and 3, but when a parallel sprue runs full the pressure at section 2 may be different. This pressure can be calculated by equating the total energies at sections 2 and 3 for the parallel sprue system. Using the indexing as shown in Fig. 9.3, we obtain for sections 1–3

$$h_1 + 0 + \frac{p_a}{\rho} = 0 + \frac{v_3^2}{2g} + \frac{p_a}{\rho} + k_2 \cdot \frac{v_2^2}{2g} + k_3 \cdot \frac{v_3^2}{2g}$$

and for sections 1–2

$$h_1 + 0 + \frac{p_a}{\rho} = h_1 - h_2 + \frac{v_2^2}{2g} + \frac{p_2}{\rho} + k_2 \cdot \frac{v_2^2}{2g}$$

Equating the right-hand sides of these equations, and because for a parallel sprue $v_2 = v_3$ (from the constancy of volume equation),

$$\frac{v_3^2}{2g} + \frac{p_a}{\rho} + k_2 \cdot \frac{v_2^2}{2g} + k_3 \cdot \frac{v_3^2}{2g} = h_1 - h_2 + \frac{v_2^2}{2g} + \frac{p_2}{\rho} + k_2 \cdot \frac{v_2^2}{2g},$$

which gives

$$\frac{p_a}{\rho} - \frac{p_2}{\rho} = h_1 - h_2 - k_3 \cdot \frac{v_3^2}{2g} \tag{9.6}$$

As the right-hand side of equation 9.6 must be positive, p_2 is smaller than p_a, or the pressure at section 2 is below the atmospheric for a parallel sprue running full. As a consequence any gas present at section 2 and surrounding the metal stream would be sucked into the stream. This phenomenon has been referred to in the technical literature as sprue aspiration of the mould gases. The conditions under which such gas aspiration can occur in gating systems are discussed in 9.9.

9.6.5 Pouring basin, sprue and runner bar system

Equating energies for this system does not involve any novel factors. Using the indexing system shown in Fig. 9.4, we obtain

$$h_1 + 0 + \frac{p_a}{\rho} = 0 + \frac{v_4^2}{2g} + \frac{p_a}{\rho} + k_2 \frac{v_2^2}{2g} + k_{3f} \frac{v_s^2}{2g} \cdot \frac{l_s}{D_s} + k_{3b} \frac{v_3^2}{2g} + k_{4f} \frac{v_4^2}{2g} \frac{l_r}{D_r}$$

Substituting for v_2 and v_3, from $a_4v_4 = a_3v_3 = a_2v_2$, leads to

$$h_1 = \frac{v_4^2}{2g}\left(1 + k_2\left[\frac{a_4}{a_2}\right]^2 + k_{3f}\left[\frac{a_4}{a_3}\right]^2\frac{l_s}{D_s} + k_{3b}\left[\frac{a_4}{a_3}\right]^2 + k_{4f}\frac{l_r}{D_r}\right)$$

where suffix f = friction
         b = bend
         s = sprue
         r = runner

From the above equation we obtain

$$v_4 = c_4\sqrt{2gh_1} \tag{9.7}$$

and

$$c_4 = \sqrt{\frac{1}{1 + k_2\left(\dfrac{a_4}{a_2}\right)^2 + k_{3f}\left(\dfrac{a_4}{a_3}\right)^2\cdot\dfrac{l_s}{D_s} + k_{3b}\left(\dfrac{a_4}{a_3}\right)^2 + k_{4f}\dfrac{l_r}{D_r}}}$$

The rate of flow at section 4 is then

$$Q_4 = a_4v_4 = a_4c_4\sqrt{2gh_1} \tag{9.8}$$

The runner bar introduces the possibility of two distinct variants in the design of gating systems. The system in which $a_4 < a_3 \leqslant a_2$ runs full throughout (if the parallel sprue is used it must be running full). This is known as the convergent or choked system and is widely used in practice. Here a choke section can be at a single location, or the whole length of the runner is choked. On the other hand, a system where $a_4 > a_3 \leqslant a_2$ is flow-controlled at section a_3, and the runner may or may not run full. This is known as the unchoked or divergent gating system, which is also widely used. The application and design of divergent systems which run full at all sections is discussed in a subsequent paragraph.

9.6.6. Pouring basin, sprue, runner bar and ingate system

When a single ingate is added to the runner bar, Fig. 9.4, the system is treated hydraulically in an analogous manner; the problem of convergency or divergency in the design of the system is also identical. An additional problem arises when several ingates are used, as in a manifold system. In a convergent system $(a_{5.1} + a_{5.2} + \ldots) < a_4 < a_3$, all the ingates will run full, but the corresponding velocities may not be equal, and consequently the flow rates through the ingates will be different. This situation is even more likely to arise with divergent systems, where there is also a possibility of some ingates not running full or even not running at all, Fig. 9.4c. Both convergent and divergent manifold ingate systems give rise to two important hydraulic problems, variation of speed in fully running ingates and the degree of filling of the ingates. Both systems are frequently used in foundry practice and are further discussed below, while their specific hydraulics are considered in more detail in the recommendations for further reading.

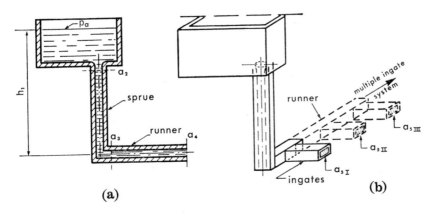

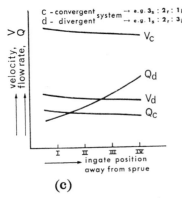

FIGURE 9.4

Hydraulics of gating systems

(a) Pouring basin, sprue and runner system.
(b) Pouring basin, sprue, runner and ingate system, with possible extension into a manifold system.
(c) Flow distribution in a multiple ingate system: convergent system, all ingates running full, and divergent system, ingates near the sprue not running full.

9.7 APPLICATION OF HYDRAULICS TO THE GATING SYSTEMS

9.7.1 Nature of the problem

The practical application of hydraulic studies to gating systems has three distinct purposes. Firstly, it is necessary to establish the non-ideal flow characteristics and numerical values of the coefficients required for the calculation of various gating systems. Secondly, these data are often obtained by using water models and therefore it becomes necessary to establish whether the results apply in the same way to liquid metals. Finally, the metallurgical significance of the hydraulic behaviour must be examined with respect to the quality of the liquid metal and the consequent properties of the casting.

9.7.2 Hydraulic relationships

The basic hydraulic relationships which apply to steady flow in simple gating systems have been dealt with in 9.6. The examples given show that, for the numerical evaluation of flow in any given gating system, it is first necessary to obtain experimentally the values of loss coefficients k, following which the other hydraulic quantities v, p and Q can be calculated for the flow conditions where the Bernoulli equations apply. The energy losses at some sections are so large that completely misleading results would be obtained if an ideal flow were assumed, for which $k = 0$ or the discharge coefficient $c = 1$. Unfortunately, hitherto, energy losses have not been determined for all the elements and combinations of elements that may be encountered in gating systems used in foundry practice. Although some data are now available, general application of hydraulic principles to the calculation of gating systems is not yet possible, but where the loss coefficients are known some practical calculations can be made.

Velocity of flow v and mould filling time t. The velocity of flow can be obtained from the Bernoulli relation for a given system, using the appropriate values for k. With the calculated value of velocity, the volume of flow Q through an ingate of cross-sectional area a can be found, since $Q = av$. The time required to fill a mould of volume V is therefore

$$t = \frac{V}{Q} = \frac{W}{\rho av} = \frac{W}{\rho ac\sqrt{2gh}} \tag{9.9}$$

where W = weight of the casting, ρ the density of the liquid metal, and other quantities are as in equations (9.3).

Various complications can arise in such a calculation if h is not constant during the mould filling. For example, in bottom gating or submerged flow, the value of h changes continuously as the mould is filled (see *Ingates Report* listed in the further reading).

Pressure. Gating systems are usually designed in such a way as to avoid negative pressures in all the gating elements. One example of negative pressure forming in the sprue system has been discussed in 9.6. The calculation of pressure in other components and at different locations of gating elements, such as the base of the sprue or entry into the runner or ingates, is carried out in the same manner as for the sprue entrance (equation 9.6), by solving the appropriate Bernoulli equations when the values of v and k have been established.

Two special aspects of pressure are the phenomenon of flow contraction, vena contracta, and the behaviour of liquid metal in the gating system of permeable mineral moulds. Examples of vena contracta are shown in Fig. 9.5.

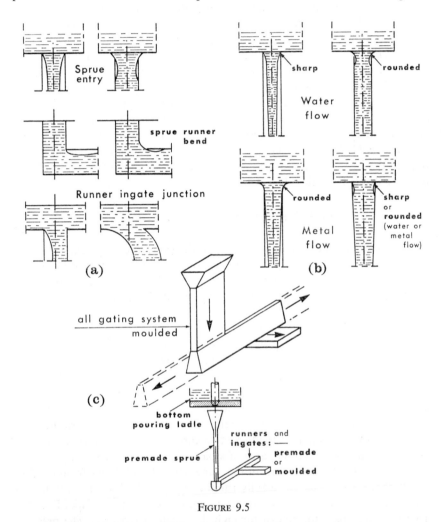

FIGURE 9.5

(a) Filling of gating components and design of entries and junctions
(b) Water and metal behaviour in sprues, as affected by sprue entry design and taper.
(c) Typical gating systems for magnesium alloys (top) and for steels (bottom).

Flow contraction occurs at those sections in a gating system where there is a sudden change in the direction of flow, for example, where the liquid enters the sprue or the runner. Consequently the problem arises of whether or not the stream will expand again and fill the next flow element fully. If the system does not run full, then the Bernoulli equation does not apply for that part of the system, whilst for a fully running system a negative pressure, if present, at the section of vena contracta would lead to gas being absorbed. The metallurgical significance of such phenomena in the behaviour of liquid metal is discussed in 9.8.

Distribution of flow in ingates. It follows from the general discussion of hydraulic principles in 9.6 that the major factors determining the distribution of flow in ingates are the magnitude of pressure and pressure distribution in the runner bar. In a convergent or choked system the pressure is higher than atmospheric and changes uniformly in the runner bar. A convergent system runs full on account of its reduced cross-section in the direction of flow, and uniformity of flow through a number of ingates can be maintained by modifications in the design of the gating elements. Considering the Bernoulli relation for flow through ingates, Fig. 9.4, we have

$$h_1 + 0 + \frac{p_a}{\rho} = 0 + \frac{v_5^2}{2g} + \frac{p_a}{\rho} + \Sigma k_i \frac{v_i^2}{2g}$$

This shows that the potential energy of the convergent system is mainly converted into kinetic energy of flow at the ingate exits, apart from the energy losses in the system. In convergent systems, therefore, the velocities of flow in ingates are high and the flow rates through multiple ingates are of a similar order of magnitude.

Divergent systems do not differ from convergent systems as far as the pouring basin and tapered sprues are concerned ($a_2 > a_1$), but in a divergent system the cross-sectional area of the runner bar is greater than that of the base of the sprue ($a_3 > a_2$). The total cross-sectional area of ingates may be still further increased in relation to the runner bar in a fully divergent system ($a_5 > a_4 > a_3$), or the ingate area may be made smaller ($a_5 < a_4 > a_3$) in a combined divergent-convergent system. The velocity and distribution of flow in ingates are different for these two cases. In a fully divergent system the increase in cross-sectional area of a part of the system is accompanied by a decrease in the speed of flow in the same part. This follows from the fact that with $a_5 > a_4 > a_3$, for a fully running system, the law of constancy of volume requires that $a_5 v_5 = a_4 v_4 = a_3 v_3$, and consequently the speed of flow from the sprue base to the exit through the ingates decreases. The pressure in the runner bar, on the other hand, increases in the direction of flow as required by the law of conservation of energy. Consequently the flow rate will be highest through the furthermost ingate, and decreases for ingates nearer to the base of the sprue, Fig. 9.4c. On the other hand, if the total area of the ingates is smaller than the runner bar area, a more uniform flow through the ingates results. A comparison of the behaviour of various ingate systems is shown in Fig. 9.4c.

Several designs of runner bar and ingates are used in practice, including gradual reduction of ingate area, or ingates located over or under the runner bar, or streamlined designs, all aiming to achieve a uniform flow through the ingates Fig. 9.2. This is an important potential field of application of hydraulics to gating systems, as at present most of such systems are designed on an empirical basis.

Energy losses. From the point of view of energy conversion, the gating system elements have two distinct objectives. The pouring basin and the sprue are normally designed for minimum energy loss and for optimum flow efficiency. Radiusing the sprue entry and tapering the sprue are the two main means towards this end. The runner bar and ingates can be designed either to accentuate the energy losses in order to reduce the overall velocity of flow (mainly by adding extra elements to the system), or to maintain low losses in the system. The latter is important from metallurgical and production considerations, as discussed in the subsequent paragraphs.

9.8 FLOW OF WATER AND FLOW OF METALS

For simplicity of experimental work most of the hydraulic studies of gating systems have been made with water models. The problem therefore arises of correlating water model data with the flow of metals in moulds.

An important hydraulic principle used in comparative hydraulic studies is that the nature of the flow is the same if the dimensionless Reynolds' Number of the flowing liquids is the same:

$$N_{\mathrm{R}} = \frac{V.D}{\mu}$$

where V is the velocity of flow, D the diameter of tubular flow, and μ the kinematic velocity, i.e. dynamic viscosity (3.2.2) divided by density. As the kinematic viscosity values of water and most metals are similar, hydraulic data based on work with a similar Reynolds' Number should hold equally well for liquid metals as for water. However, for flow in gating systems, which is considered in this chapter, some important differences in the two types of liquid do arise.

With most typical gating system designs and materials parallel sprues run full with water but not with liquid metals, Fig. 9.5. This is due to the difference in wetting characteristics of the two types of liquid. Consequently, it is essential to use tapered sprues for liquid metal flow in mineral moulds if an efficient flow is wanted, otherwise only the head of the metal in the pouring basin is used in the energy conversion. A parallel sprue could, however, be made to run full with liquid metals by introducing a convergent element in the distribution part of the gating system (choked systems). These differences in wetting behaviour also apply to runner bar and ingates in divergent systems.

The difference in wetting, and hence the degree of filling, with water and liquid metals leads to another problem, that of pressure distribution. As indicated, negative pressure can result in some gating elements, such as the entry into a parallel fully running sprue. As parallel sprues do not run full with metals, a negative pressure at the sprue entry cannot obtain in this case. If, on the other hand, a critically tapered sprue is used, or a parallel sprue made to run full by choking, the pressure at the sprue entry will not be below atmospheric. The problem of gas aspiration is therefore very much less frequent with liquid metals than with water systems, but gases can be entrapped by the metal stream through some other cause, as discussed below.

9.9 METALLURGICAL AND PRODUCTION SIGNIFICANCE OF GATING HYDRAULICS

An examination of the present empirically developed gating practice for mineral mould processes reveals striking variations in the design of gating elements ranging from those used for magnesium base alloys to those used for steel Fig. 9.5c. Such variations stem from both metallurgical and production requirements. Alloys which readily form a solid oxide, particularly that of the film type, at the free surface of the liquid are sensitive to conditions of flow, and particularly to turbulence in the gating system. Consequently, for alloys based on or containing aluminium, magnesium, chromium or similar metals, gating systems are used which have the least turbulent flow and therefore the smallest flow speed at the ingates. For alloys which hold oxygen in solution (e.g. many copper, iron or nickel-base alloys) gating systems based on fully choked elements are applied. With the former turbulent flow results in the formation and trapping of oxide films, and these find their way into the structure of the casting (inclusions). With the latter, on the other hand, this danger is relatively small, and some of the advantages of choked but more turbulent systems can be utilised (e.g., faster filling of the system, less danger from slag entrapment and higher casting yield).

Apart from the behaviour of oxygen, metallurgical factors include considerations of speed of flow as a cause of mould erosion, and sand entrapment in casting, the control of occlusion of slag and flux particles in liquid metals, feeding behaviour (Chapter 10), and the nature and kinetics of reaction between the liquid metal and the mould face, known as metal-mould reaction, 8.10.1. With steel castings, for example, Fig. 9.5, the gating system is frequently moulded in special minerals or with special binders to reduce mould erosion (exogenous inclusions). Special mould dressings and mould gas reaction inhibitors are used with some magnesium, aluminium and copper base alloys.

Production requirements often conflict with hydraulic or metallurgical requirements, since they demand the simplest gating system from the point of view of moulding, fettling the casting, and casting yield (11.5.1). Divergent systems, because of their design, give lower casting yields than do convergent systems. The main production considerations are the choice of a particular gating system and the determination of its dimensions. The choice of a system can be resolved in terms of the hydraulic, metallurgical and production principles discussed in this chapter. Flow conditions consistent with metal behaviour and its liquid properties, meeting optimum production requirements and leading to the production of high quality castings, can be satisfactorily achieved in most cases. The crucial problem is the dimensioning of gating systems, which is at present solved by certain empirical formulae (discussed by Ruddle), or by experience. There is little doubt that such formulae could be replaced with great advantage by methods based on applied hydraulics, once the required hydraulic data of gating systems have been obtained experimentally. The economic incentive for such research depends largely on the requirements in properties of castings. So long as castings are used in the industry as "unavoidably" having some process defects, then a large number of possible gating design variations can be used to obtain what are loosely defined as acceptable castings. It is only when high standards in castings are required that optimum gating systems become essential.

9.10 SPECIAL GATING SYSTEMS

Among those gating systems that differ from the typical systems discussed in this chapter, the most interesting is that of pressure die casting, Fig. 8.2.4. The Bernoulli equation is reduced to

$$\frac{p}{\rho} - \Sigma \text{ losses} = \frac{v^2}{2g}$$

However, the importance of energy evaluations is superseded here by the importance of the actual flow conditions determined by the design of the gating elements. Flow conditions are responsible for the occlusion of mould gases, distribution of flow stream in the die, and die cavitation phenomena. With the high velocity of metal flow, the metal stream occludes mould atmosphere gases much more readily than in simple gravity pouring. Furthermore, at localised positions pressure conditions in the metal stream go beyond those predicted by the Bernoulli relation. Localised low pressure regions form, which cause the liquid to collapse into them, and this type of cavitation phenomenon leads to rapid die erosion.

9.11 SUMMARY

Design and dimensioning of gating elements in the production of shaped castings is one of the more important fields of survival of empirical methods and skill in founding. Unlike many of the problems of liquid state and nucleation, for example, where experimental methods are inherently difficult, the present low state of development of applied gating hydraulics is largely due to lack of experimental effort. And yet, in spite of this situation, even such simple considerations of hydraulics as introduced in this text are of considerable practical interest. The principles discussed not only allow a much better understanding of current empirical practices, but also make possible a large number of improvements in the design of gating elements and in devising and applying methods of casting quality control.

FURTHER READING

RUDDLE, R. W., *The Running and Gating of Castings*, Institute of Metals, London, 1956.
Symposium on the Principles of Gating, American Foundrymen's Society, 1953.
GRUBE, K. and EASTWOOD, L. W., A study of the principles of gating, *Trans. Amer. Found. Soc.*, 1950, **58**, 76.
JEANCOLAS, M., COHEN, G. and HANF, H., Hydrodynamic study of horizontal gating systems, *Trans. Amer. Found. Soc.*, 1962, **70**, 503.
Ingates Report, *Proc. Inst. Br. Foundrymen*, 1955, **48A**, 306.

CHAPTER 10

FEEDING AND THE SOUNDNESS OF CAST METALS

10.1 INTRODUCTION

As a result of the decrease in specific volume (shrinkage) of the liquid and during the freezing of most alloys, and because of the characteristic features of crystal growth surfaces, cavities of various sizes, shapes and distribution appear in the structure of cast metals unless extra liquid metal is available at all surfaces where liquid to solid transformation is taking place. The feeding of castings is defined as the process of compensating for volumetric contraction in such a way as to ensure that 'sound' castings, free from or with a controlled amount of shrinkage cavities, result. Unlike the problem of metal flow into a mould cavity, casting soundness cannot be readily interpreted in terms of a single basic theory. This is because the feeding of castings involves problems of both engineering and metallurgical origin, and often considerations in the field of production engineering are also included. The engineering aspects are related to casting design and to the features of the mould construction that may affect the production of sound castings. For example, a large size sphere free from cavities could not be cast in a long freezing range alloy which contracts in volume during freezing when use is made of any of the conventional mineral mould processes discussed in Chapter 8. It is equally obvious from metallurgical considerations (Chapter 6) that an alloy freezing with a smooth solidification front would be more conducive to achieving a high soundness in all castings.

Feeding of castings is a good example of a casting problem where often practical experience alone can supply an industrially acceptable solution, quite apart from numerical and engineering evaluations. Although the former method is more generally used for dealing with the feeder problem in foundries today, in order to maintain the general theme of this book this problem too will be considered mainly in terms of scientific principles. However, in order to establish a basis for a more scientific approach, the general conditions of feeding castings will first be discussed.

10.2 PRINCIPLES OF FEEDING

10.2.1 Engineering

The functional requirements of cast metals are seldom such that their design cannot be partly modified to help the feeding problem. In general,

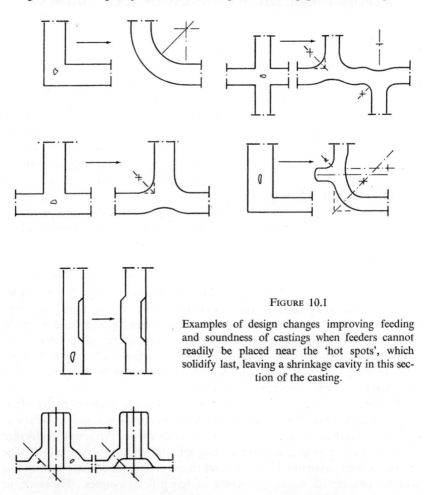

FIGURE 10.1

Examples of design changes improving feeding and soundness of castings when feeders cannot readily be placed near the 'hot spots', which solidify last, leaving a shrinkage cavity in this section of the casting.

the solution of the feeding problem begins at the drawing board, and metallurgical and production principles are then jointly examined. The major engineering principles are illustrated in Fig. 10.1. Hot spots should be avoided, together with sections which in the course of freezing are cut off from the liquid necessary to compensate for the volumetric contraction. Tapered sections can be used to avoid feeding difficulties along central planes, and gradual rather than abrupt changes in casting sections should be arranged. Rigidity and strength should be achieved by reinforcements whenever possible rather than by section thickening. The foundry engineering features of pattern and mould design must also be considered.

10.2.2 Metallurgy

Feeding problem variables include the magnitude of the liquid and freezing contractions, the nature of zones and fronts of crystal growth (Fig. 6.9), the nucleation of cavities, the spatial distribution and direction of the solidification

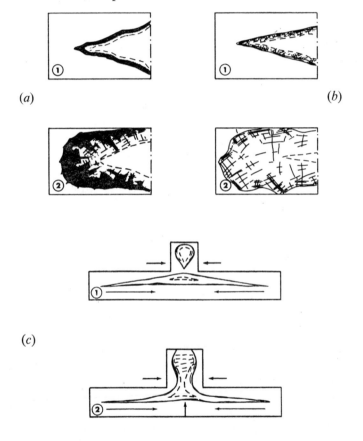

FIGURE 10.2 Metallurgical factors of feeding

(a) Nature of the solidification front: (1) continuous (smooth) and (2) discontinuous (rugged).
(b) Depth of the solidification zone: (1) shallow and (2) deep.
(c) Direction of freezing: (1) mainly horizontal, feeding liquid cut off,; (2) horizontal and vertical, feeding liquid in continuous contact with the solidifying liquid.

front, i.e. temperature gradients (Fig. 10.2), the feeding pressure and the composition of the feeding liquid. The total volumetric contraction is mainly dependent on alloy composition, pouring temperature and mould rigidity. The optimum solidification conditions for feeding are flat solidification fronts or thin or narrow solidification zones (Chapter 6), and a freezing direction towards the liquid metal reservoirs or feeders attached to an external surface of the casting (directional feeding), Fig. 10.2. Feeding pressure can be controlled to a marginal extent and the alloy composition determines the physical properties of the liquid relevant to feeding.

10.2.3 Production engineering

A feeding system designed on engineering and metallurgical principles must also take into account the production factors of the casting process; these include the moulds (methods of moulding and mould properties), ways and means of feeder removal (fettling castings at the end of the process), costs of feeding (materials and process) and finally acceptance tolerances in properties and the nature of the permissible volume of cavities or porosity left in the casting (Chapter 11).

10.3 THEORETICAL APPROACH TO THE FEEDING PROBLEM

Given a mould cavity of a specific shape, contained in a mould material of known thermal properties, can the feeding problem be solved numerically to obtain a casting of a given degree of soundness with an alloy of known physical properties? This is the kind of problem which cannot be readily solved by calculation unless experimental work is carried out first to establish the conditions under which a possible theoretical solution could apply. The present position in foundry practice of arriving at the feeder solution by trial and error methods is largely a consequence of the lack of such prior experimental data. It is clear from the discussion of the freezing mechanism (Chapter 6) that a numerical solution of the feeding problem would have to take into account the following considerations: (a) type of growth morphology (i.e. continuous skin or pasty mesh), (b) growth direction and magnitude of temperature gradients, (c) the dependence of growth morphology on the alloy composition and casting variables (pouring temperature, vibration and pressure), (d) nucleation characteristics of cavities and hence their total volume in relation to the total amount of volumetric contraction and (e) the effect of casting geometry (shape and volume), and the thermal properties of the alloy and the mould on the growth interface and temperature gradient variables. Clearly, the feeding problem involves too many variables for a general numerical solution to be found which would apply to all alloys and all casting processes. Instead a numerical solution of feeding presupposes, a priori, a selection of materials and of freezing conditions for which the feeding process variables can be interrelated. For example, one such convenient group would be: type of alloy (i.e. growth interface type) and a specific casting process (heat transfer variables). Here the required temperature gradients and feeder volumes could be interrelated to produce castings of a given soundness. In general, therefore, different types of alloys and casting processes give rise to specific feeding problem solutions.

10.3.1 Feeding mechanisms, feeding conditions and feeder calculation methods

Different feeding mechanisms can be analysed with reference to Fig. 10.2. Two separate cases will be considered: a closed feeding and an open feeding system, Fig. 10.2c, together with a single (continuous), Fig. 10.2a(1), and a double (discontinuous) growth interface, Fig. 10.2a(2), in each case. Solidification will be considered to proceed away from the mould wall, and a temperature gradient, however small or large, exists during the freezing. Such a temperature distribution is nearest to the condition of freezing of castings in practice, and the condition of isothermal solidification throughout the casting can be interpreted as an extreme case of gradient solidification.

In an open system with a single and continuous growth interface, characteristic of single temperature freezing alloys, the growing solid front compensates its freezing contraction directly from the adjacent liquid. The overall volumetric contraction for the amount of solid formed manifests itself in the gravity sinking of the top liquid level at the highest point, i.e. in the feeder. A combination of atmospheric and metallostatic pressures, jointly with a progressive temperature gradient solidification towards the feeder, lead to an overall and final shrinkage cavity being left in the feeder, assumed to be of adequate volume. The pressures exerted also contribute to the skin of the casting in the early stages of its growth being pressed against the mould wall. For the case of three-axial solidification of a casting (X-length, Y-thickness, Z-width), the condition of full feeding can therefore be expressed by the relation:

$$\frac{\Delta t}{\Delta X} > 0, \quad \frac{\Delta t}{\Delta Y} > 0, \quad \frac{\Delta t}{\Delta Z} > 0$$

where ΔX, ΔY, ΔZ represent small but finite increases in thickness of the solid layer frozen per unit time, and Δt the temperature gradient at the interface of this grown layer.

In a closed system, all other conditions being the same as in the previous example, the solid skin forms all round, including the top level of the feeder, so that the remaining liquid is eventually cut off from the atmospheric pressure. As the skin continues to grow, the free volume space created by the feeding of the solid formed can lead to the following two physical situations: either the liquid contained inside the casting skin fractures to create a free volume (pore, shrinkage cavity) to release the negative pressure which otherwise would form, or alternatively the solid skin collapses from outside to relieve such a negative pressure. Two different feeding situations, or a combination of the two, can therefore arise in a closed system:

(a) The total volumetric contraction of freezing is taken up by the solid collapse from outside (this situation is further analysed in the recommendations for further reading).

(b) Volumetric contraction leads to the formation of a cavity which is finally left in the feeder, or in both the feeder and the casting if the feeder neck freezes prior to the casting.

In industrial casting practice the freezing conditions often lead to a combination of (a) and (b). Nucleation of cavities occurs in this case by one or more of the following mechanisms: puncturing of the casting skin by the atmospheric pressure, gas bubbles being first generated in the liquid and tearing up of the liquid, usually at a solid–liquid interface.

The mathematical statement of feeding in a closed system is identical with that for the open system, but the conditions for full feeding and retention of the original dimensions of the cast shape demand that a cavity should also form in the feeder. The advantage and importance of open system feeding in industrial practice is thus readily apparent.

The case of alloys freezing over a temperature interval which results in a discontinuous growth interface, Fig. 10.2a(2), presents a more complicated feeding problem. The end of the solidification interface demarcating all the solid region, which is at the solidus isothermal, t_S, is fed from the adjacent liquid, and so are all the other surfaces of the growing crystals whose front ends are sticking out into the liquid demarcating the onset of the solidification front (the liquidus isothermal, t_L). In between these two growth fronts and the corresponding temperature isothermals, i.e. in the pasty zone, the axial growth directions of the crystals are random in space, so that crystal surfaces (e.g., arms of dendrites) may interconnect, thus cutting off some pockets of the liquid from the main liquid. The whole freezing system can thus be split into a large number of small and independent freezing systems, each depending on its own feeding liquid. Clearly therefore full feeding can only take place in the pasty zone if no fully isolated liquid pockets form as a result of crystal growth, i.e., if the liquid is continuous throughout the pasty zone. Alternatively, each liquid pocket finally either results in a small cavity or is 'fed' by the solid collapse mechanism (identical to case (a) of closed systems). In an extreme example of slow cooling and long freezing range alloys, liquid pockets, some of which are interconnected, can extend from the centre to the outer surface of the casting. The presence of atmospheric pressure in this case helps in the formation of contraction cavities by relieving the negative pressure which would otherwise form in the liquid inside, no puncturing of of the casting skin being necessary for cavity nucleation.

Feeding requirements for freezing range alloys can therefore be expressed by two statements:

$$(a) \quad \frac{\Delta t}{\Delta X(Y, Z)} > 0$$

$$(b) \quad \frac{(t_S - t_L)_X}{X_S - X_L} \gg 0, \quad \frac{(t_S - t_L)_Y}{Y_S - Y_L} \gg 0, \quad \frac{(t_S - t_L)_Z}{Z_S - Z_L} \gg 0$$

where t_S and t_L are the actual solidus and liquidus temperatures of the alloy, and X_S, X_L, etc. are distances separating these temperatures in the casting during freezing. The former expression is identical with the feeding condition for open systems of single temperature freezing alloys and states the need for a temperature gradient towards the feeder. Part of the overall contraction of long freezing range alloys can therefore be compensated from the feeder. The second expression is equally important and states the condition for the feeding in the pasty zone itself, which demands a sharp temperature gradient in this zone, i.e. that the depth of the pasty zone in all the three directions of freezing should be as shallow as possible. If this zone is deep (a shallow temperature gradient) only a small amount of feeding from the feeder occurs and the bulk of the volumetric freezing contraction is left in the casting in the form of dispersed and mainly discontinuous porosity. Feeding of long freezing range alloys thus requires not only directional freezing but also very steep temperature gradients, particularly in the direction of the Y and Z axes. The statement of the conditions of feeding of open or closed systems for long freezing range alloys follows along similar lines to that for single freezing interface alloys. Cavities are nucleated in the same way as with the short freezing range alloys, except that the air suction through the outer skin is much more favourable with the long freezing range alloys.

It follows from this discussion of feeding mechanisms and feeding conditions that there are four basic elements which control the feeding process: the nature of the growth interface, the magnitude of the temperature gradients, cavity nucleation factors and the design of the feeding reservoirs or feeders. A calculation approach to the solution of the feeding problem for a given alloy could therefore be along two different lines: either by calculation of the required temperature gradients or, alternatively, by calculating the size of feeders on the assumption that their proper evaluation and design implicitly includes the need for and the existence of required temperature gradients. As the temperature gradient approach is much more complex, most solutions of the feeding problem used at present in practice are based on feeder calculations alone. As full feeding implies absence of cavities, the factor of cavity nucleation does not come into feeder calculation methods.

Furthermore, it is clear that for production of a fully sound casting the choice of alloy, and the design of casting, mould and feeder have to be taken into consideration. However, in practice, the free choice of alloy or casting or mould design is restricted by engineering and production considerations. The biggest freedom in technical selection lies in the design of feeders, but even here there are some restricting factors. Consequently, very often in practice a compromise solution is accepted, which allows a certain amount of porosity to be left in castings, consistent with their application in service.

10.3.2 Heat transfer approach

The solution of the feeding problem using the heat transfer approach can be attempted in the following manner. The mould cavity contains molten alloy, the heat content of which is Q cal. This heat is being gradually transferred through the mould–metal interface into the mould body and subsequently to the surroundings. If heat transfer laws could be applied to calculate the solidification time, t_c, of a casting in such a manner that the liquid would solidify last at an external surface, then the dimensions of a feeder could be calculated so that it would solidify after the casting, $t_f > t_c$. The solution of the feeding problem by heat transfer theory assumes, therefore, that feeding conditions can be stated in terms of solidification times of castings and feeders. This approach does not take into account numerically the temperature gradients in the casting or the freezing mechanism of the alloy.

Cooling molten metal in a batch mould is an example of an unsteady state of heat transfer. During pouring the mould interface at temperature θ_0 is heated to θ_i, and a varying temperature gradient is established in the mould wall, so that in the direction X the temperature θ_x will be continuously changing with time t. Considering a flat mould face, Fig. 8.11, with heat transfer in the direction X only, the change of temperature θ with time t and with the distance x from the interface is given by the equation

$$\frac{\partial \theta}{\partial t} = \alpha \cdot \frac{\partial^2 \theta}{\partial x^2} \tag{10.1}$$

where α is a temperature diffusivity factor. The general quantitatial solution of this equation to obtain temperature θ distribution in a mould (distance x) as a function of time t is cumbersome, but still possible when the necessary thermal data and boundary conditions are known. A simplification can be introduced for feeder calculations where the unsteady interface temperature θ_i behaves approximately as a constant. This condition is closely met when cooling pure metals or eutectics in sand moulds. With this approximation and the other boundary conditions known, equation (10.1) can be integrated to obtain the temperature distribution in the mould as a function of time and distance from the interface. Using the same terms as previously and assuming

θ_i to be constant and reached immediately after filling the mould, the temperature θ_x after a time t and at the perpendicular distance x from the interface is given by

$$\theta_x = \theta_0 + (\theta_i - \theta_0)\,\text{erfc}\left(\frac{x}{2\sqrt{\alpha t}}\right) \tag{10.2}$$

where erfc $\dfrac{x}{2\sqrt{\alpha t}}$ is an error function. (The origin and methods of application of this are discussed in the recommendations for further reading.) The importance of the temperature distribution given by equation (10.2) is that it can be used to obtain the amount of heat transferred through the interface by finding the rate of change of temperature θ with distance x, namely $(\partial\theta/\partial x)_{x=0}$, and by substituting the general heat transfer equation through a surface:

$$\frac{\partial Q}{\partial t} = -k\left(\frac{\partial\theta}{\partial x}\right)_{x=0} \tag{10.3}$$

The substitution of the differential of (10.2) into (10.3) and integration lead to

$$Q = \frac{2k(\theta_i - \theta_0)\sqrt{t}}{\sqrt{\pi\alpha}} \tag{10.4}$$

where Q = heat transferred/unit surface area.

The term α (temperature diffusivity) can be replaced by a term defined as heat diffusivity β:

$$\alpha = \frac{k}{\rho c} \quad \text{and} \quad \beta = \sqrt{k\rho c}$$

where k = thermal conductivity,
  ρ = density,
  c = specific heat of the mould material.

Equation (10.4) then becomes

$$Q = 1\cdot128\beta(\theta_i - \theta_0)\sqrt{t} \tag{10.5}$$

For a given set of experimental conditions, metal, mould and temperature factors become constant and equation (10.5) can be further simplified into

$$Q = c_1\sqrt{t} \tag{10.6}$$

The value of c_1, the 'mould constant', can be obtained from thermal data and used to compare the cooling capacity or chilling power of different moulding materials. Because the value of θ_i is a function of the freezing temperature of the alloy, the value of the chilling power changes with the alloy composition. Table 10.1, gives the values of the chilling power for the casting temperature range of aluminium and copper alloys. The values of the chilling power of metal moulds cannot be obtained in the same way owing to the unsteady behavious of θ_i.

TABLE 10.1

Moulding method	Interface temperature, °C	Chilling power cal/cm^2/min $\frac{1}{2}$
Plaster of Paris	550	80
	1080	180
Green sand	550	110
	1080	260
Dry sand	550	90
	1080	240
Magnesite	550	230
	1080	350

As the bulk of heat transferred results from the freezing of the liquid metal contained in the mould, it follows directly from equation (10.6) that the thickness D of the solid plate formed by the freezing process is equal to

$$D = c_2\sqrt{t} \tag{10.7}$$

Equation (10.7) is known as the square root–time relation for solidification of metals in mineral moulds, and it can be used when the conditions used for its derivation are satisfied and the value of the constant c_2, the 'solidification constant', is known.

For example, with the values of c_1 given in Table 10.1, the solidification times of flat plates (if the end effects are neglected) can be calculated. This follows from equations (10.5), (10.6) and (10.7), which show that

$$c_2 = \frac{c_1}{L\rho'}$$

where L is the latent heat of fusion and ρ' is the density of the cast metal. Unfortunately, the complex geometry of most castings in practice and the uncertainty of the heat transfer boundary conditions in general, together with lack of thermal data, make it impossible to obtain the solidification time of castings in this way. However, the value of the chilling power is useful when considering directional cooling (10.4.2) and the relative heat extraction powers of various mould materials, insulating pads and chills.

The square root–time relation is important in the theoretical study of the problem of feeding castings, as it clearly indicates that two distinct further developments of the heat transfer approach to the feeding problem are possible. Either equation (10.1) can be solved to deal with the geometrical factors, such as shape and design of castings for the various boundary conditions encountered, or alternatively equation (10.7) can be applied. The first involves mathematical difficulties, although these are not so formidable with the availability of modern computer techniques. In the second case, although equation (10.7) was originally derived for a limited set of conditions, by using various approximations and correction factors, its application can be extended to the general solution of the feeding problem, based on the heat transfer approach. This approximation method is the basis of several feeder dimensioning theories in use at present, and these are summarised in the following paragraphs.

10.3.3 Numerical and graphical methods

The usefulness of equation (10.7), derived for the solidification of a semi-infinite plate, can be further extended. The thickness D of the solidified plate (i.e. casting skin) can be expressed in terms of its volume V and surface S from the geometrical relation

$$D = \frac{V}{S}$$

By substituting into equation (10.7) we obtain

$$t = c\left(\frac{V}{S}\right)^2 \tag{10.8}$$

Equation (10.8), known as Chvorinov's rule, is the fundamental equation of approximate feeder calculation methods based on the heat transfer approach. It was Chvorinov who first empirically observed that the solidification times of castings can be related to their volume-to-surface ratios. Equation (10.8) simply means that the geometrical restrictions used in deriving equation (10.7) do not hinder its application in practice provided that the main boundary conditions hold, and this is generally so for slowly cooled castings in mineral moulds, Fig. 10.3.

Amongst a number of attempts to apply equation (10.8) for feeder calculation the best known is that of Namur. The solidification time of the feeder is equated to that of the casting, using equation (10.8), and by introducing several simplifications the solution proposed by Namur becomes

$$V_f = k_1 f_1 \left(\frac{V}{S}\right)_c + k_2 f_2(V_c) \tag{10.9}$$

where V_f is the volume of the feeder, k_1 and k_2 are metallurgical and thermal constants, and $f_1\left(\dfrac{V}{S}\right)_c$ and $f_2(V_c)$ are functions of the volume and surfaces of the casting. Because of the difficulties in evaluating these functions, Koppe has proposed a graphical solution which is discussed in the paper listed in the recommendations for further reading.

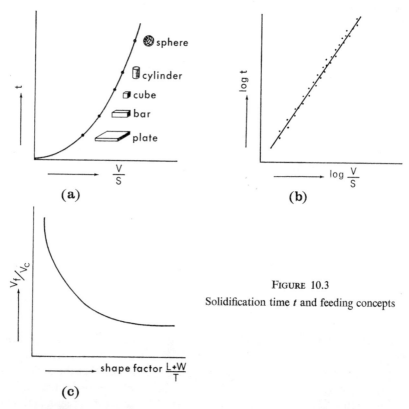

(a)

(b)

(c)

FIGURE 10.3

Solidification time t and feeding concepts

(a) Solidification time for simple geometry castings of equal casting volume but different surface areas.

(b) Empirical relations obtained by Chvorinov for castings of varied volumes and surfaces cast in mineral moulds.

(c) Geometry method of feeder dimensioning (Pellini, Bishop).

10.3.4   Geometrical methods

The scarcity of thermal and metallurgical data required for the application of equation (10.9) had led a number of investigators (Pellini, Bishop and others; see further reading) to simplify further equation (10.8) and obtain empirical graphical relations between the feeder to casting volume ratio

V_f/V_c and the shape factor of the casting expressed in terms of length L, width W and thickness T of a casting, Fig. 10.3. When graphs of this kind are obtained experimentally for a given alloy and type of mould, the main problem is to express the geometry of a complex design casting in terms of its shape factor. This difficulty has been examined by Wlodawer and Jeancolas, who propose feeder size evaluation methods retaining the basic relationship between the feeder dimensions D_f and the volume to surface ratio of the casting $\left(\dfrac{V}{S}\right)_c$

$$D_f = k \left(\frac{V}{S}\right)_c$$

These workers propose special, and in some ways differing, methods of obtaining the values for $\left(\dfrac{V}{S}\right)_c$, as well as experimental methods for determining the value of k (see further reading).

The feature common to all feeder evaluation methods based on the heat transfer approach is the implicit assumption that the feeder design alone can be used to resolve most, if not all, of the feeding requirements of a casting. This explains why these methods are more successful with short freezing range alloys, and why they are only partly successful, or fail altogether, with the long freezing range alloys, where the transverse temperature gradients are more important than the longitudinal ones for effective feeding.

10.3.5 Experimental methods

The major obstacle to the numerical calculations of feeder dimensions previously described is the difficulty of applying heat transfer laws to complex casting designs and their numerous variations, combined with the anisotropic and varied nature of the moulding materials used in many casting processes and the lack of thermal data in general. Several empirical methods have been proposed instead for feeder dimensioning, using geometrical and thermal data relationships obtained through direct experimental measurements for a given alloy. A method of this type is that proposed by Weston, Tiryakioglu and the author, Fig. 10.4.

If the solidification times t_f of typical feeders are obtained by direct cooling curve measurements, then for a casting having the solidification time t_c the correct size of feeder can be read directly from the feeder time graphs, using the relationship

$$t_f = \lambda t_c \tag{10.10}$$

The value of λ varies for a given alloy with the solidification time and the casting volume and is determined experimentally, Fig. 10.4c. This method therefore depends entirely on the use of quantities which have been obtained or verified experimentally. This approach is further considered in 10.5.7.

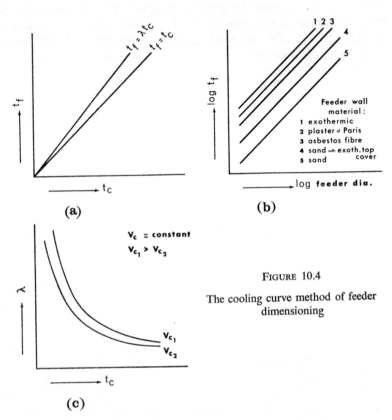

FIGURE 10.4

The cooling curve method of feeder dimensioning

(a) Solidification time of feeder t_f exceeds that of casting t_c by a factor λ.

(b) Curves for a particular feeder design (plotted from experimental data), enabling correct feeder dimensions to be read off for a known casting solidification time.

(c) The conversion factor λ; variation with solidification time of casting, for castings of constant volume, is determined experimentally.

10.4 EMPIRICAL APPROACH TO FEEDING

10.4.1 General definition of the problem

A survey of the methods of feeding currently used in foundries shows that none of the proposed feeder calculations (10.3.3 and 10.3.4) has been successfully applied in practice generally, but partial applications have been made. As some kind of solution to the feeding problem is inseparable from the day to day production of castings, it is important to understand and assess the current practice of feeding, and to appreciate the difficulties of any attempt to deal with the problem of feeding on a more scientific engineering basis. For the analysis of current feeding practice it is convenient to consider separately the type of casting process, the composition of the alloy and the methods of feeding used.

10.4.2 The casting process and the direction of cooling

The mould materials and the method of casting are the two major variables that determine the rate and direction of cooling. In any particular

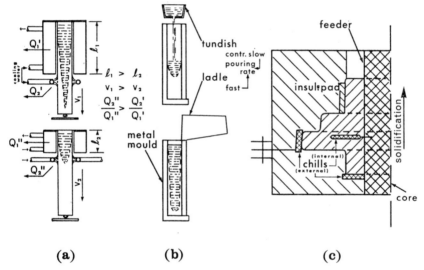

(a) **(b)** **(c)**

FIGURE 10.5 Examples of casting process variables and their effects on feeding
(a) The effects of mould length l, casting speed v and cooling method variation Q_2/Q_2 on feeding conditions in continuous casting.
(b) Pouring rate and feeding conditions in batch ingots (small size) and gravity moulds.
(c) External aids: chills and insulating pads in mineral moulds.

process these two variables can be modified to achieve more favourable feeding conditions. The cooling rate in parts of a casting is generally altered in such a way that the direction of cooling, or freezing of the alloy, is made to proceed towards the feeder.

The feeder size, number and location for a given casting have to be evaluated before the start of the moulding process, and most of the empirical work on feeding has been concerned with this problem.

Examples given in Fig. 10.5 illustrate some methods of achieving directional cooling in practice. In continuous casting of ingots the rate and direction of cooling are controlled by the design of the mould and the cooling system, but both of these variables can be further modified in a given system by the speed and method of ingot removal from the mould, the rate of flow of the cooling water, mould lubrication, and metal pouring temperature. In a batch ingot mould, too, the mould material and design control the cooling conditions, but further variations in cooling rate and direction can be obtained by changing the pouring rate and the mould and liquid metal temperatures. The same holds with minor modifications for gravity and pressure die castings. In gravity die casting, for example, directional cooling is dependent on mould design, mould materials and cooling methods and on the application of mould dressings of different thermal properties. In general, the degree of flexibility in varying the cooling rate and direction is less with sand moulds than with metal mould processes. Most mineral mould materials can be considered as thermal insulators, and only minor changes in cooling conditions are obtained by varying pouring rates and metal and mould temperatures. This is the main reason why with mineral moulds directional cooling involves to a higher degree the consideration of casting design, gating system, use of external and internal cooling agents or chills, and numerous variations in feeder design and materials (10.4.4 and Fig. 10.7).

In any given casting process an attempt is made to co-ordinate the effects of the variables controlling the cooling rate and direction to obtain feeding conditions consistent with the achievement of other desirable properties. It is clear from Fig. 10.5 that in continuous casting the feeding process is also continuous and dependent on the shape and depth of the solidification zone. In small ingot batch moulds similar conditions can be achieved by controlling the pouring rate and other casting variables. When this is not possible, an external feeder has to be used. With sand moulds, on the other hand, an external feeder is essential unless volumetric contraction is so small that feeding can be neglected, or the solid contraction from the outer surface takes up all the volumetric shrinkage (10.3.1).

10.4.3 Alloy composition

The two main alloy composition variables affecting feeding behaviour are the magnitude of the volumetric contraction in the liquid state and during freezing, and the freezing temperature range. With some alloys, such as certain grey irons, feeding is required mainly to compensate for the liquid contraction ΔV_L, and the freezing contraction ΔV_F is very small or zero. For a given alloy and volume of casting the minimum volume of the feeder $V_{f\,min.}$ is therefore

$$V_{f\,min.} = \Delta V_L + \Delta V_F$$

or in terms of coefficients of contraction, ξ_L and ξ_F, and the volume of the casting V_c

$$V_{f\,min.} = \xi_L V_c + \xi_F V_c \qquad (10.11)$$

In most casting processes it is impossible to arrange for solidification to proceed in such a way that the feeder solidifies last while at the same time its volume is just large enough to compensate the total contraction of the casting ($\Delta V_L + \Delta V_F$), with the last residual liquid in the feeder being taken up by the casting and leaving a flat surface where it joins the casting. In other words, the feeder must have a volume greater than $V_{fmin.}$, and this excess of metal is cut off from the casting.

$$V_f > V_{fmin.}$$
$$V_f = c \cdot V_{fmin.} \tag{10.12}$$

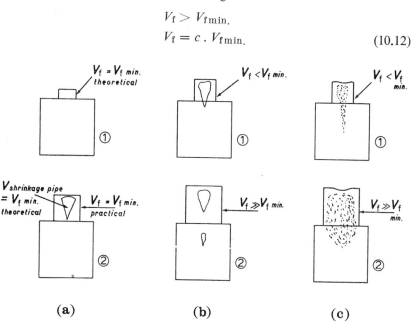

FIGURE 10.6 Feeder volume and alloy composition

(a) Optimum feeder volume (1) theoretical and (2) practical.
(b) (1) undersize and (2) oversize feeders, with short freezing range alloys.
(c) (1) undersize and (2) oversize feeders with long freezing range alloys.

The practical problem is therefore to determine the value of the coefficient c, which is not constant but varies with alloy composition, volume and design of the casting, method of feeding, and to a lesser extent with other factors, as shown in Fig. 10.6.

It is clear from Fig. 10.6 that there is an optimum volume of feeder, which is just adequate to feed a part or the whole of the casting. An undersized feeder leads to central or external shrinkage cavities with short freezing range alloys, or to dispersed cavities with alloys freezing over a solidification interval. An oversized feeder may not eliminate such cavities, since the feeder is part of the casting and therefore influences the solidification of the casting section to which it is attached. Consequently, with an oversized feeder a secondary pipe may form in the casting section with short freezing range alloys. Also shrinkage cavities may be made worse rather than better when the pasty solidification zone of the feeder is extended into the casting with long freezing range alloys. In addition, oversized feeders may increase the hydrostatic pressure through the liquid metal on the mould walls and this may lead to mould cavity dilation. Expansion of the mould cavity in this way, late in the progress of solidification of the casting, can lead to increased shrinkage cavity volume and an oversize casting. The best size for a feeder cannot therefore be determined without considering the casting process, alloy composition and feeder design variables.

10.4.4 Feeding methods

With a given casting process, casting design and specified alloy, several feeding methods may be possible. Variations include feeder shape, location, material and number, and also pressure and other external aids, Fig. 10.7. It follows from equation (10.8) that the shape of the feeder giving the longest solidification time is that of a body with the minimum surface for a given volume, i.e. a sphere. As this shape is not convenient for moulding, the most frequently used feeder shapes are cylindrical or semi-spherical. Feeder location is largely controlled by the casting design, and whether top, side or overlap feeder is used depends also on moulding and fettling factors. From the point of view of removing the feeder or fettling the casting, the area of contact of the top feeder with the casting can be reduced by tapering the neck of the feeder, or by insert cores, giving knock-off feeders. The top area of a feeder may be exposed to the atmosphere, or a feeder may be completely enclosed in a mould and is then termed a blind feeder.

The walls of the feeder may be of the same material as the rest of the mould, but its solidification time can be extended by using sleeves of special materials for insulation or for extra heat generation. Insulating materials are now available for feeder application for temperatures up to 1,600°C. Additional heat can be supplied to the metal in the feeder either by using special exo-thermic mixtures for feeder walls (usually premade sleeves) or by electric energy, normally an electric arc. Exothermic mixtures are based on the well known reactions:

$$Al + Fe_2O_3 \rightarrow Al_2O_3 + \Delta H \text{ cal} \quad \text{and}$$

$$2\,Al + 3O_2 \rightarrow 2\,Al_2O_3\, \Delta H_1 \text{ cal}$$

the kinetics and heat quantities of which are controlled by the choice and mixing of the neutral fillers and reactants.

The number of feeders may be varied from one feeder serving several castings in a mould to one casting requiring several feeders, each possibly differing in design and other details. In addition to the casting design and moulding methods, an important factor is the feeding range or distance,

Fig. 10.7. Feeding range is defined in practice as the length of the casting away from the feeder that can be satisfactorily fed in a casting of given geometry and alloy composition, contained in a given mould material. Full theoretical evaluation of the feeding distance is still not possible, but for

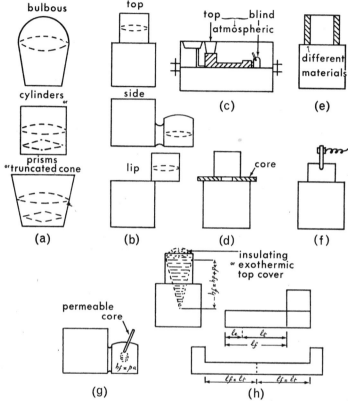

FIGURE 10.7 Elements of feeders and feeding methods

(*a*) shapes, (*b*) position on the casting, (*c*) location in the mould, (*d*) knocking off, (*e*) materials (sand, insulators, exothermics or combinations of these), (*f*) heating, (*g*) pressure, and (*h*) feeding distance.

several alloys the values of the feeding distances l_f have been obtained empirically for different alloys, casting designs and moulding conditions. These are usually expressed in terms of an end or corner factor l_e and a length factor l_t, both of which are related to the thickness t for bar and plate castings:

$$l_f = l_e \cdot t + l_t \cdot t$$

For feeding a steel bar of thickness t made in a green sand mould, for example, the feeding distance has been experimentally obtained as

$$l_f = 6\sqrt{t}$$

The values of feeding distances obtained experimentally for different alloys and casting dimensions are given in publications listed at the end of the chapter. If the required feeding distances are available, the total number of feeders required for a given casting can be determined.

An important factor in the consideration of the feeding problem is pressure. In most foundry processes using gravity pouring, the total feeding pressure, h_f, is made up of two components, the hydrostatic pressure h_g and the atmospheric pressure h_a. The total pressure per unit area at the solidification front in the mould depth h is

$$h_f = h_g + h_a = h\rho' + p_a$$

As soon as the top surface of the feeder has solidified, the residual liquid is fully enclosed by the solidified casting skin, and the feeding pressure is only

$$h_f = h\rho'$$

Consequently in most feeding methods an attempt is made to keep the feeder top surface open and accessible to the atmospheric perssure by using special insulating or exothermic top covers, and for blind feeders a permeable sand core insert can be used, Fig. 10.7. Special considerations of pressure in centrifugal and pressure die casting are discussed in 10.5.

10.4.5   Inter-relationships of the feeding variables

From the general discussion on the feeding variables in casting practice it is clear that for any given group of main variables—casting process, alloy composition and casting design—there are a large number of dependent variables, and a number of feeding methods may be possible in any given problem. For any fixed combination of independent and dependent variables it should be possible to arrive at a method of evaluation of the optimum feeder size by selecting and correlating suitable parameters the numerical value of which has been obtained empirically. Some empirical methods discussed in the further reading are based on such an approach. Others are based on heat transfer relationships (10.3), but their use depends on the values of empirical coefficients for practical applications.

10.5   CURRENT FEEDING PRACTICES

10.5.1   Ingots

From the feeding point of view, the continuous ingot casting process, Fig. 8.1, is one of the best solutions of the feeding problem applicable to simple shapes. By varying the cooling conditions and mould design, it is possible to obtain a favourable solidification direction and to reduce the depth of the pasty zone. Simple ignot shapes can therefore be cast by this process as sound as those in the wrought condition. As the casting process is continuous or semi-continuous, the problem of an external feeder reservoir does not arise, but at the very end of the process a short end length of the ingot is usually less dense and is therefore discarded.

The cooling, and hence the feeding, conditions in the continuous process are controlled by the heat transfer properties of the mould material, the mould design, the method of ingot removal (oscillating moulds, or stop–go ingot withdrawal), the heat transfer properties of the alloy being cast, and the heat transfer conditions of the cooling medium both within and below the mould. The ratio of the heat transferred horizontally within the mould to that transferred vertically below the mould controls the shape and depth of the solidification front, Fig. 10.5a. By varying the mould materials, the length of the mould, the speed or method of ingot withdrawal, and the amount of cooling water, a wide range of feeding conditions can be obtained. Other metallurgical factors also have to be considered, such as ingot surface quality, structure of the alloy and internal stresses. Because of such complex and often conflicting requirements it is generally not possible to arrive at the optimum cooling conditions for continuous casting from heat transfer calculations only. However, heat transfer calculations can be applied to predict the effects of numerous process variables affecting the cooling conditions of a given ingot, once the boundary conditions have been established.

Feeding conditions characteristic of some batch ingots are shown in Fig. 10.5b. Within a limited range of small size ingots it is possible to obtain a favourable narrow pasty zone by controlling the pouring rate or the rate of rise of the cooling water level, as in the tin can mould. In the cases shown in Figs. 8.3b and 10.5b, as with the continuously cast ingot, the problem of an external feeder does not arise. With large size non-ferrous ingots or with cooling conditions encountered with steel ingots poured into cast iron moulds, on the other hand, it is not often possible to dispense with an external feeder. With batch-cast steel ingots two main principles are employed, Fig. 10.8. For low carbon steels ($C < 0.2\%$) the degree of deoxidation may be controlled so that sufficient oxygen is left in solution to promote and control the extent of the rimming reaction:

$$C + O \rightarrow CO/CO_2$$

The volume of gas bubbles trapped in the ingot during solidification is controlled so that it equals the total volume of cavities that would otherwise have resulted through liquid and freezing contractions, and consequently an ingot with a flat top or head results. Such ingots are known as rimming, rising or balanced steel ingots. With still larger carbon contents, balancing such a reaction is difficult for practical purposes; for this reason the molten steel is fully deoxidised or killed, and hence it contracts in volume in the usual way and requires feeding. A wide variety of feeder designs and materials can be used for killed steels, varying from the—now largely historical—hot top feeder in fireclay, sand or similar material, to feeders lined with more modern insulating and exothermic materials, Fig. 10.8. The correct feeder

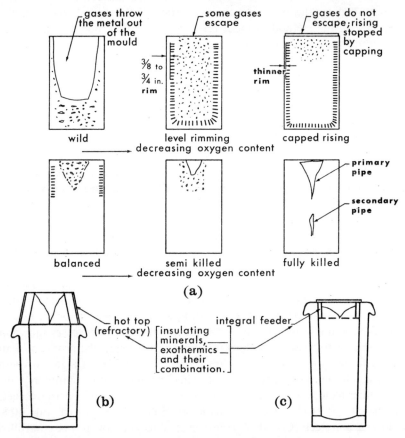

FIGURE 10.8 Rimming gas reaction and feeding phenomena of carbon steel ingots

(*a*) The effect of decreasing oxygen content on rimming and the formation of rim and shrinkage cavities as a result of rimming gas reaction.
(*b*) 'Classical' hot top (dozzle) method of feeding killed steel ingots.
(*c*) Integral (built in) feeder of more recent design.

size and shape is arrived at experimentally by considering mould variables (shape, wall taper, wall thickness and mould temperature), metal variables (pouring temperature and rate) and heat conservation factors in the feeder (relative heat losses upwards, sideways and downwards). These variables can be related empirically for feeder size evaluation and control purposes.

The extent of feeding in relation to ingot soundness is only one of several metallurgical quality requirements, so that empirical heat transfer relations are usually used as an auxiliary rather than a fundamental approach to feeding. Similarly, such empirical data can be usefully applied to determine time of ingot stripping from the moulds, and hence to the conservation of soaking pit heat energy prior to the hot working of ingots.

10.5.2 Gravity die (permanent) mould castings

In principle, apart from the complication of casting shape and design, the problems of feeding gravity die castings and batch ingots are identical. However, in comparison with many ingots (>1 ton), gravity die castings are generally lightweight (<2 cwt), and feeding problems are solved on the basis of a controlled gating system, supplemented by suitable casting and mould design, and its heating and cooling arrangements.

A casting is designed firstly for favourable directional freezing, Fig. 10.1, and secondly for convenience of making and operating the mould. As with ingots, the best mould working temperature is a compromise between the optima for feeding the castings, for surface quality, dimensional requirements, for cast structure and stress behaviour. The surface temperature of the die, although ideally constant, in production differs from one part of the mould to another. Such temperature differences help to promote directional freezing. Correct balance is achieved by proper mould design, by adjusting the gating system to control the flow of the metal into the die, by adjusting the thickness and conductivity of mould dressings, and by cooling and heating of the parts of the mould, Fig. 8.2. In this way it is possible to obtain progressive freezing towards the feeder, which thus becomes an integral part of the casting. Heat conservation in the feeder by special insulation and heat generation is used in gravity casting as an additional measure, but it somewhat complicates the process.

When casting design factors and alloy characteristics are not favourable to directional freezing, a small amount of dissolved gas can be intentionally left in the alloy. Some of this is rejected during solidification and reduces the extent of volumetric contraction by substituting finely dispersed gas cavities (a similar idea to the rimming action, but on a much smaller scale). The cavities formed have less detrimental effects on the properties required in some castings than do contraction cavities. In general, feeders for gravity die castings are not at present designed on the basis of either heat transfer or empirical calculations.

10.5.3 Pressure die castings

Static and gravity forces characteristic of feeding conditions described in previous examples are augmented in pressure die casting by a considerable external pressure. It might be expected therefore that sounder castings would result, as the liquid alloy is forced during solidification into the cavities formed by volumetric contraction. However, the fundamental technical aim of this process is to obtain a casting of high dimensional accuracy and very good surface quality. These features are obtained with a partial sacrifice of soundness and structural characteristics. As discussed in Chapter 9, when a mould is filled at high pressure, the liquid metal contains an appreciable amount of occluded mould cavity gases. These gases reduce the requirements for feeding, but in most cases they remain as residual gas cavities in the castings.

The principles of casting, mould and gating design, and of mould and metal temperature control, apply to pressure die castings in the same way as to gravity die castings. In most instances special feeders are unnecessary, but sometimes feeder pads have to be used as extensions to the casting in spite of the occluded gas, Fig. 8.2(5). At present feeders are designed on the basis of experience, and neither heat transfer nor empirical calculations are yet used in pressure die casting practice. The growing demand for pressure die castings with better mechanical properties is gradually leading to changes in the mould filling process (vacuum and injection pressure process modifications), and this is combined with improvements in casting soundness.

10.5.4 Centrifugal casting

Centrifugal casting occupies a very special position among casting processes. Metal or mineral moulds can be used and either ingots or shaped castings can be produced. The feeding problem in all these cases is very simple. The liquid metal is introduced into the mould in such a way and at such a rate that the direction of freezing is towards the rotation axis or towards the feeder, Fig. 8.3. A narrow freezing zone obtained by the control of process variables, combined with a high pressure on the liquid metal from the centrifugal force, gives castings which are in most cases nearly as sound as wrought metal products.

10.5.5 Sand castings

Gravity feeding of mineral moulds is by far the most difficult feeding problem encountered in all foundry practice. This is due not only to the wide variations encountered in mould designs, casting shapes and weights, and alloys, but also—and this is the major factor—to the very slow rate of heat abstraction characteristic of mineral moulds. Consequently the temperature gradient from the surface to the centre at any cross-section of the casting is very small or almost zero at the time of freezing. Similarly, the temperature gradient from a given section towards the feeder is equally small. This makes it difficult, if not impossible in some cases, to obtain freezing towards the feeder or narrow freezing zones favourable for feeding. With short freezing range alloys, where the freezing zone is shallow or absent (Fig. 10.2), a high degree of soundness can often still be achieved. With long freezing range alloys, when the freezing zone extends across most of the cross-section, full feeding is in many cases impossible. The solution of feeding problems with sand castings begins, as in previous examples, with a consideration of possible improvements in the casting design. Following this, gating system variables, such as the distribution and number of ingates, and the location and type of feeders, are arranged in such a way as to promote temperature gradients which, though small, are essential for feeding. Finally, the temperature gradients, which are the key factor in feeding, are also controlled whenever possible by applying external or internal chills, supplemented by insulating pads and section tapering, Fig. 10.5.c.

If enlargement of mould cavity volume occurs during solidification, due to liquid metal pressure, this increases the amount of liquid metal required for feeding. Consequently mould design, sand compacting and moulding box locking must be examined with a view to increasing the stability of the mould cavity volume or mould rigidity. This factor is important, particularly with alloys which show a tendency to outward pressure in freezing, as some grey irons and bronzes.

10.5.6   Feeders designed on the basis of practical experience

Sand foundry practice today relies largely on skill, supported in some cases by empirical and heat transfer methods, for solving feeding problems. This involves an assessment of the various factors discussed in this chapter for arriving at an optimum solution of feeding. The question therefore arises whether a formal theoretical method based on temperature gradients or heat transfer methods would prove advantageous in the long run for calculating the feeder sizes and feeding conditions. As indicated in Section 10.3, this is a question in which the short-term but high-cost basic research which would be necessary has to be balanced against the long-term benefits of consistent production of adequately sound castings. An example of the kind of experimental work that is required is considered in the following paragraph.

10.5.7   The cooling curve method

A number of empirical methods originally based on the heat transfer approach and supported by different measurements have been proposed for calculating feeder dimensions for shaped castings made in mineral moulds (10.3.3 and 10.3.4). The aim is to obtain optimum dimensions for a selected combination of casting and feeding conditions. The methods differ in the selection of the parameters which serve as dependent variables. According to Chvorinov's equation (eq. 10.8), the parameter selected most widely is the volume to surface ratio V/S. However, the complexity of casting shapes and designs makes it necessary to introduce various simplifications and assumptions for determining the value of the effective cooling surfaces as distinct from the total casting surface. In addition, the variation in mould design and materials casting processes, alloys and feeding methods complicates a general solution for a feeder evaluation method in terms of the value of V/S.

The cooling curve method, Fig. 10.4, eliminates some of the difficulties encountered with the V/S parameter. The solidification time of feeders for various alloy groups (taking account of other variables such as size, shape, design and material) can be determined by normal cooling curve recording techniques. After determining the solidification time of various castings in the same way, the optimum feeder size can be obtained from the feeder graphs, using the concept of the optimum feeding factor λ (eq. 10.10). The value of λ gives that ratio of feeder to casting solidification time which results in a casting of a defined degree of soundness. For a given alloy, the numerical value of λ varies with the values of casting volume V_c and its volume to surface ratio. The successful application of the method depends, therefore, on determining experimentally the value of λ. The results so far obtained for some aluminium base alloys have demonstrated that the cooling curve method can be successfully applied to feeder evaluation problems in sand foundry practice. The extension of the method to other alloys and other casting processes depends on the availability of values for λ and simultaneous determination of its full significance and its relation to temperature gradients and feeding conditions.

The practical application of this method is very simple. For a given casting the number, type, location and wall material of feeders, and the means for achieving directional freezing can be selected as the basis of factors discussed in this chapter. Finally, the size of the optimum feeder has to be determined. The volume of the casting being known, the value of λ is obtained from the appropriate λ/V_c graph, Fig. 10.4c, and the value of t_f can then be determined (eq. 10.10). The feeder diameter is read directly from the graph (Fig. 10.4) and its volume checked from equation 10.11.

The general usefulness of the cooling curve method is that it implies a need for an adequate number of experimental data on the temperature conditions of feeding, which in the long run will provide the background for a more general solution of the feeding problem.

10.6 SUMMARY

There is little doubt that the soundness of castings, which depends largely on achieving proper feeding conditions during solidification, is one of the major factors in assessing the structural quality of cast products, particularly for their applications in service where mechanical properties are of prime importance. Although a general numerical solution of the feeder problem is too difficult in terms of temperature gradients or heat transfer laws, some simplified solutions based on these laws and supported by direct measurements are plausible. Since the data necessary for the application of such simplified solutions have not yet been made generally available, present foundry practice largely relies on the application of skill, experience and recorded and correlated data obtained in production.

FURTHER READING

RUDDLE, R. W., *Solidification of Castings*, 2nd ed., Institute of Metals, 1957.
CHVORINOV, N., Control of solidification of castings by calculation, *Proc. Inst. Br. Foundrymen*, 1938–9, **32**, 229.
NAMUR, R., Einfluss Zone der Steiger, *Giesserei, Tech-wiss. Beihefte*, 1958(20), 1077.
PATTERSON, W. and KOPPE, W., Ein allgemeines Verfahren der Steigerbemessung, *Giesserei, Tech-wiss. Beihefte*, 1960(30), 1647.
WLODAWER, R., *Gelenkte Erstarrung von Stahlguss*, Giesserei Verlag, 1961.
JEANCOLAS, M., CHEVRIO, R. and VIROLLE, X., Méthode générale pour la détermination des masselottes, *Fonderie*, 1964(215), 1.
BISHOP, H. F., MYSKOVSKI, E. T. and PELLINI, W. S., A simplified method for determining riser dimensions, *Trans. Amer. Found. Soc.*, 1955, **63**, 271.
WALLACE, J., *Fundamentals of Risering Steel Castings*, Steel Founders Society, 1960.
CAMPBELL, J., *Origin of Porosity in Cast Metals*, Thesis, University of Birmingham, 1967.

CHAPTER 11

QUALITY CONTROL AND TESTING THE PROPERTIES OF CASTINGS

11.1 INTRODUCTION

The economic objective of a casting process is to produce the required number of castings with the specified properties at a minimum overall cost. The failure or success of a modern production casting process often depends on the testing and quality control methods used. Some of the required properties directly related to the general cast structure of alloys have been discussed in Chapter 7. These and other functional properties of castings will be considered in this chapter from the metallurgical point of view of the various methods used for testing and quality control in production.

It is clear from the characteristics of the casting processes discussed in previous chapters that variations in the properties of a casting are likely to arise even when a single type of casting is produced repeatedly in the same alloy and by the same process. The object of testing and quality control in production is to attain only such variation in properties of castings as is within the required tolerances for any particular casting process or combination of processes. Some tests are concerned more specifically with production engineering problems and fall outside the scope of the present text.

The metallurgically important properties of finished castings could be described as five sisters, as they all begin with the letter 's': surface, size, structure, soundness and stress. Casting quality depends on the optimum being obtained in one or more or all of these properties, and variations from the optimum are related directly to some aspects of metal or mould behaviour and indirectly to the numerous process variables. As optimum properties cannot always be obtained in production, an agreed 'standard' level is usually taken for the basis of quality control. It is convenient therefore to divide the testing and control methods in foundries into two groups: those carried out during the process itself and those concerned with the evaluation of the properties of the finished castings.

11.2 PROCESS QUALITY CONTROL: METAL TESTS

11.2.1 Composition

A frequent variable affecting metal behaviour in a casting process is the alloy composition. Variations in this arise mainly through compositional variations in the raw materials used and compositional changes occurring in the melting and the casting processes (Chapter 2). Physical and chemical methods which can be used for testing the composition of metals in foundries are summarised in Table 11.1. The analytical methods differ essentially in their accuracy, the time required for the analysis, their special applicability for the determination of specific elements, and the overall cost of the analysis. Frequently, the requirements of a foundry demand the application of two or more different analytical methods. Furthermore, some analytical methods are interdependent, and one method may follow another in the analysis of a single alloy.

Assessing metal composition usually involves analysis of the composition of the solid charge and/or of the composition during melting and prior to pouring. Methods of sampling, number of samples and frequency of analysis depend on whether the analysis is for process control or for standard specifications. Samples taken just prior to pouring are normally taken to be compositionally identical with the finished castings, and are acceptable as specification analyses, but in some cases specification analysis samples are taken from the actual castings produced. Some approximate methods of process composition control have been summarised in Chapter 3.

11.2.2 Standard specification and process control analyses

The properties required in the casting normally determine the degree of tolerance in the final alloy composition. Standard specifications for cast metals are agreed upon by various trade, professional and government bodies. They are arrived at after consideration of the casting process and castings applications, and summarise levels and deviation tolerances in composition and often in other properties. They establish a system of alloy identification which regulates the relationship between producer and consumer, as done in other branches of industry and commerce. In the simplest case standard specifications give only the composition and tolerances in the main alloying elements. The continuous demand for improved castings with higher consistency, has led to a continuous revision in standards. For some applications, the standards may at present specify the tolerances in unwanted or 'tramp' or 'residual' elements, the presence of which, if not exceeding the limits specified, may not affect deleteriously the required properties. Some tramp elements or impurities are particularly injurious and have to be kept to very much finer limits. In addition to the compositional standards, a number of other tests and property specifications (particularly pertaining to the gaseous elements) may also be required (11.2.4).

TABLE 11.1 *Principal quantitative methods of analysis of alloys*

General method	Specific method	Principle	Characteristic	Applications
Chemical and physico-chemical	Gravimetric (precipitation)	Element or its compound separated by precipitation.	Normal accuracy 0·5% to 1% of element content, dependent on the technique; chemically simple, but time consuming, requires skill; also used for standardising	Classical general method for most alloys, ores or slags, routine or standards or refereeing
	Gravimetric (electrolytic)	Element or its compound separated from solution by electrolysis	Simple and fast, but with limited applications; accuracy as above or better	Mainly for Cu, Ni, Cd, Zn, Pb, routine or standards or refereeing
	Gravimetric (combustion)	Element separated from solid alloy by forming gaseous product with another gas at high temperature	Routine, fast, but very limited; accuracy as precipitation	Mainly for carbon in alloys, routine or standards or refereeing
	Volumetric (titrometric)	Calibrated reagent solution added to solution containing unknown element and 'end point' determined using different indicators	A number of variants based on different chemical principles; faster but generally equally or less accurate than gravimetric, depending on element, method and technique	Classical general method, more widely used than gravimetric. Routine or standards or refereeing
	Volumetric (colorimetric)	Colour of solution containing unknown element evaluated by different methods against calibrated standard	Fast, instrumented, requires standards, but of less general application than titrometric; accuracy less than 1%	Mainly for lower concentrations and impurities (light alloys and transition elements), routine or standard specifications

TABLE 11.1 (*contd.*)

Chemical and physico-chemical	Polarographic	Electrolysis of solution with cathode of falling mercury	Instrumented, fast, useful for specific elements, usually non-ferrous or Pb, and Cu in steels; accuracy approx. 1%	Mainly for trace elements, complementary to other methods, routine or standard specifications
Physical	Spectrographic	Characteristic radiation spectrum of elements, obtained by sparking the sample, dispersed and recorded; intensity compared against standards; vacuum and neutral gas sparking extends range of elements, and direct intensity evaluation (quantometer) raises speed of analyses	Variants in sparking diffraction and intensity evaluation; highly instrumented, fast, less dependent on operator's skill than chemical methods; accuracy 2% to 5%, improves with heavier elements; high initial cost	Repetition, routine, standard specifications, or low alloy contents; more suitable for heavier elements
	X-ray fluorescent	Secondary X-ray spectrum of solid or solution sample	Highly instrumented and high initial cost; accuracy as spectrographic	Complementary to spectrography but for larger alloy contents

Process control analysis methods used by the producer for the control of quality may be identical with or in some cases additional to those required for specification analyses.

11.2.3 Composition and properties of castings

It is clear that variations in property requirements of castings lead to a wide variation in compositional standards and their controls. Compositional standards are continually changing with new knowledge of the effects of alloying elements, tramp elements and impurities on the properties of castings. An interesting metallurgical problem arises in this field in that certain elements, most frequently of gaseous origin and not specified in the standards, may be particularly harmful to the properties of castings. The reasons for this anomaly are partly physico-chemical and partly metallurgical, as such elements cannot readily be determined analytically as a routine test. Consequently, composition standards demand chemical analysis of specified elements only and—as a safeguard against possible effects of 'unspecified' elements—other property tests are normally added to the standards tests. Such property tests can be used either for process control (quality tests) or for standard acceptance tests. The same tests can be used for both purposes, but often separate ones are used.

11.2.4   Casting 'quality' tests

While composition standards establish the basic chemical identity of an alloy, it is clear from Chapter 6 that a wide range of structures, and hence different properties, can be obtained in castings made in an alloy of the same composition. Furthermore such properties can be affected by structural features, such as cavities, which are partly independent of the composition. It is therefore essential to carry out various property tests and controls to supplement the composition quality and acceptance testing. Such property tests may include the effects of gaseous or other elements not normally determined by the analyses, but their main objective is to deal with the structural and specific properties behaviour of the alloy as affected by composition, casting and heat-treatment process variables. Some of the tests used for this purpose have already been considered in Chapter 4 (gas tests) and Chapter 7 (structure tests). The most common among the other property tests are those for assessing mechanical properties (11.2.5) (structure sensitive properties).

11.2.5   Test bars and mechanical properties

Test bars of various designs, and cast by various methods, either serve to establish specific and fundamental property values of cast metals, such as mechanical and physical properties, or they are used for process quality control or for standard acceptance test purposes. Test bars used for evaluation of fundamental cast metal properties can be cast in a laboratory or in a foundry and the designs, methods and casting techniques used vary with the specific objectives of the tests. Similarly, test bar practice for process quality control varies according to the specific objective of the control in question. Standard specification test bars, on the other hand, are used for the evaluation of specified mechanical or physical property standards, supplementing, or sometimes even replacing, chemical composition acceptance tests. Frequently, test bars used for the determination of fundamental properties, or for process or acceptance control, are identical in design and are cast in identical manner.

Some typical examples of standard tensile test bars used in foundry processes are shown in Fig. 11.1. The bars are small castings, the design and casting techniques of which show wide variations. The shape (e.g. round, rectangular or 'cast to shape'), horizontal, vertical, single or multiple castings, and whether the test pieces are cast separately or attached to the casting, are all arbitrary choices, largely based on experience, but 'standardised' by industrial agreements. From a production and industrial point of view, tensile test bar practice fulfils an obvious practical need, not only in providing data on mechanical properties of cast metals, but also for process quality and acceptance control of finished or heat-treated castings. The basic assumption of both quality control and standard acceptance practice is that the test

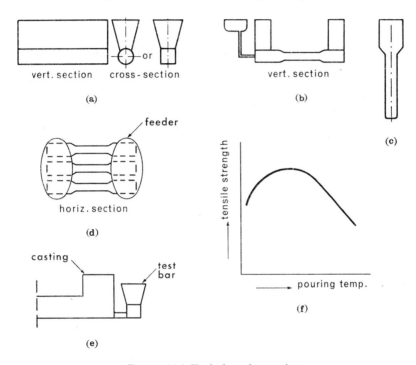

FIGURE 11.1 Typical test bar castings

(a) Wedge type (for some brasses and steels).
(b) Cast to shape (bronzes and malleable irons).
(c) Vertical (fed for light alloys, unfed for cast irons).
(d) Multiple (bronzes and some light alloys).
(e) Attached test bar.
(f) Variation in tensile properties of test bars with pouring temperature (some alloys).

bars poured from a melt separately, but at the same time as the production castings, represent the 'quality' of the metal poured. Mechanical properties attained in actual castings may or may not differ from the corresponding properties of the test bars, and the properties or quality of finished castings have to be determined separately (11.5).

However, there are certain aspects of some standard test bar practices that can be directly misleading. For example, if the test bar represents the quality of the metal prior to pouring, then the casting process of the test bar itself should not be a source of variation in test bar properties. In present practice, however, test bar casting technique can be varied so widely that the test results can be completely erroneous as to the original 'quality' of the metal. Another very serious weakness of most of the present test bar industrial practices is the uncertainty of the correlation of the properties of the test bar with those of the finished castings.

Cast test bar practice epitomises scientific uncertainty regarding the mechanical behaviour of cast metals compared with wrought metals. Mechanical properties are structure-sensitive and in the cast state this is reflected in a wide range of variation in structure. Consequently there is an urgent need for fundamental studies of the properties of cast metals in relation to their structure. With the wrought metals, on the other hand, structural conditions can be controlled with greater precision and the corresponding properties have been, in most cases, determined. From the point of view of production and commerce, cast test bar practice meets most of the requirements of process quality control provided that the metallurgical principles are applied in casting the test bars and the results obtained in testing are interpreted in a metallurgically sound manner. In many instances of using test bars for process quality control, statistical methods of result analysis have proved very useful (11.4).

11.2.6 Other tests and properties

In addition to the tensile test bar properties, other mechanical properties, such as hardness, impact resistance and ductility in bending, or some physical properties, can be tested to obtain fundamental data or for process quality or standard acceptance controls. Similarly, a variety of metal quality control tests can be used in a casting process, either with the liquid metal (e.g., pouring temperature) or with specially cast samples (e.g., macro- and microstructure, fluidity, fracture and gas content). The object of all such tests is the quality control of specific properties of finished castings (11.5). Most of these tests have been developed empirically and as a rule their full conversion into a more scientific test is difficult or impossible.

The simplest example of a process quality control and acceptance system of alloys in some foundries is that based entirely on alloy composition tests. The more complex case is met in foundries where a number of composition tests as well as property tests are carried out which may require correlation and interpretation (11.4). In contrast, with some alloy groups, e.g., malleable and grey irons, composition tests are carried out mainly for process quality control, while various property tests are used for both process control and for acceptance purposes. The reason for this is that with these alloys the required acceptance properties can be obtained within a range of compositional variations.

11.3 MOULDS: PROCESS CONTROL TESTS

Unlike the metal tests, which are primarily concerned with metal quality and to a smaller extent with property problems, mould tests and the resultant properties are mainly concerned with process and production quality control, but often these properties have a direct bearing on the casting quality. For example, sand testing practice (8.9) is directly concerned with process and production problems of mould making in sand founding (mould quality control) but the results of sand properties tests can often be correlated with the casting quality. Mould quality control tests include various tests on raw materials used for mould making and tests on prepared moulding mixtures or prepared moulds. Such tests differ from one casting process to another, and often vary within a single casting process depending on the degree of quality control required, Chapter 8. Sand testing practice in foundries varies a good deal in the method and frequency of sampling, and particularly in the application of test data (11.4). In metal mould casting processes, for example, the tests cover the properties and applications of mould dressings and include measurements and control of the cooling system of the mould, such as its temperature and the flow rates of cooling media.

11.4 APPLICATION OF TEST DATA: SYSTEMS OF QUALITY CONTROL

Test data are recorded on process, metallurgical or production control cards or charts, and these have to be analysed and applied for the purpose of quality control. The methods vary with their specific objectives, for example, maintenance and reproducibility of quality of castings in production, analysis and elimination of causes of casting defects, and quality improvement of castings.

The simplest system of quality control is that of preparing tables or graphs of process or standard test data on a time or production schedule basis, which reveal trends or deviations from the accepted standards. Remedial action is often directly apparent or can be assessed on the basis of previous experience. Sometimes, however, the explanation of the causes of defective castings cannot be readily obtained directly from record cards or plotted graphs, when it may be valuable to determine whether this is pure chance or whether it is a significant factor. If the record data are adequate, they can be analysed to establish the variable which is the most probable cause of the deviations. The method of analysis in these instances is based on applied statistics.

Applied statistics make use of the science of probability and variations to determine whether a series of recorded events may reveal a pattern which could explain the behaviour and some of the causes of the events. Applied statistics in a foundry control system is not a substitute for direct metallurgical interpretation of certain tests, but rather an additional tool for revealing where and what action may be necessary. An example of the application of statistics to chemical analysis data is shown in Fig. 11.2. The chemical composition of alloying elements A and B, as determined by chemical analysis, are graphed as a record of melting events, such as charge-melted, addition of temper alloys, etc. If such graphs are kept on a time basis for a number of consecutive melts it may not be readily apparent whether there is a tendency for A or B to systematically decrease or increase above the expected values. The results can be replotted, for a given number of melts, as gains or losses above the calculated values. The second graph shows that there is a systematic gain of element B, but the element A shows no definite tendency of variation. Thus, the composition of the charge for element B can be modified without resorting to any further statistical analyses. However, in the case of element A, it is possible by applying statistical methods to show whether there is a gain or loss tendency not shown in the graphs. For both elements A and B a further useful control graph can be obtained by plotting the composition and indicating maximum and minimum specification analyses as well as the standard deviations of results. Methods of statistical analysis are discussed in detail in the recommendations for further reading.

11.5 TESTING AND EVALUATION OF THE PROPERTIES OF CASTINGS

11.5.1 Principles of evaluation

Metallurgically an ideal casting is that which possesses optimum properties and is produced with the highest casting yield. The casting yield is defined as the per cent ratio of the weight of the finished product to the weight of the poured casting. It is impossible to obtain 100% casting yield, as some metal must be used for the gating and feeding systems; often also for trimming or machining some of the casting surfaces. The production casting yield is the ratio of the total weight of finished castings of required acceptable properties to the total weight of metal melted. The production yield is usually smaller than the casting yield, because of the inevitable production of some castings the properties of which fall outside the acceptance limit. The main object of production control systems is to aim that the production yield approaches the

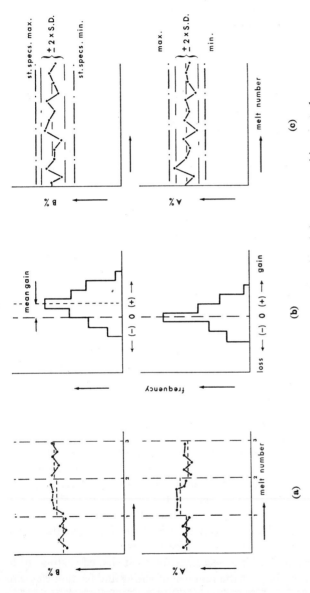

FIGURE 11.2 An example of the application of statistics to composition control of alloys in casting processes

(a) Direct plotting of alloying element content for the various stages of a melt and for several melts.
(b) Replotting the results of graph (a) in terms of frequency of losses or gains in the alloying element content for a given number of melts.
(c) Replotting the results of (a) in terms of standard deviation (S.D.), also indicating standard specification for the composition.

casting yield as closely as possible. The production yield in foundries varies usually between 50 and 90%. The production yield figure is a good indication of the success or failure of quality control methods used in foundries.

A new casting process goes into full production when its metallurgical, engineering, economic and production characteristics have been fully evaluated. An additional aim of the metallurgical and production control of surface, size, structure, soundness, stress and other requirements is that of achieving the highest production yield. Normally, no additional examination or evaluation of properties of castings is carried out beyond the scope of the quality control systems previously discussed. A casting or production yield study, however, often requires a detailed examination of the causes of rejected castings. This is an essential part in the development stage of a casting process, and is regularly carried out in foundries handling a variety of casting designs or alloys or casting processes. It is therefore of some importance to analyse the main factors arising in the examination of the properties of rejected castings or process scrap.

11.5.2 Surface of castings

The surface requirements of castings vary widely with their applications. Ingots for rolling require as a rule a better surface quality than ingots for extrusion. In the former case it is desirable that the original cast surface quality of the ingot should give an acceptable surface quality of the final wrought product, but this is seldom achieved, and scalping of the ingot surface, i.e., machining or grinding off approx. 0·25 in. prior to or after hot-working is a frequent practice, particularly with non-ferrous ingots. Extrusion billets, on the other hand, leave most of their ingot surface blemishes on the residual ingot skull retained in the container of the extrusion press, and surface quality tolerances of cast extrusion ingots are very wide. The surface quality requirements of shaped castings also vary, and the degree of smoothness and freedom from surface imperfections is frequently dependent on the functional or aesthetic tolerance and the influence of surface protection and finishing coatings which may be applied to castings (painting, enamelling, spraying, plating). For many engineering applications, an arbitrary surface quality standard is agreed upon between the user and the producer of castings. Surface quality requirements of cast metals thus raise two problems: evaluation of surface quality combined with detection of imperfections and the metallurgical interpretation of the causes of surface defects.

Prior to surface evaluation, it is necessary for most shaped castings to be fettled. The term fettling implies the removal of gating and feeding elements and any other surplus metal resulting in the process, followed by surface cleaning. Surface cleaning is a major production factor in the casting process when using mineral moulds. The difficulties of surface cleaning increase with the pouring temperature of alloys and the size of castings. A variety of cleaning methods can be used depending on cast surface condition, the hardness of the alloy and on the surface smoothness required. They include degreasing, pickling, vapour blasting, shot blasting by hydraulic or pneumatic means, and barrelling. Surface cleaning and examination may be necessary prior to or after heat treatment. Similarly, certain parts of the casting have to be machined (11.5.3) and defects may become apparent only after machining.

Casting surface smoothness measurement, which is not practiced widely, can be carried out using optical, electrical and mechanical methods and comparing with the arbitrarily agreed standards of surface quality.

The simplest method for detection of surface defects is that of visual examination, or using small magnifications ($< \times 10$). Most macro-size surface defects are revealed in this way. Imperfections on a still finer, or micro-scale, are revealed by one of the several non-destructive methods of surface examination (Fig. 11.3). Those most widely applicable in casting practice are variants of the penetrant methods, based on the principle of a suitable organic liquid or dye penetrating into the surface cavity imperfections of immersed or swabbed castings, and the dye seeping back on to the casting surface after removal of the bulk of the liquid. The principle of the magnetic flux method, which can be applied to magnetic alloys, is to produce a local disturbance in flow lines (magnetic flux) of fine magnetic particles applied to the surface of the casting, either dry or suspended in a liquid. Magnetisation can be achieved by permanent magnets or by current induction, and subsurface defects up to a depth of 0·05 in. can also be revealed. The eddy current method, used for magnetic and non-magnetic alloys, utilises the principle of measuring the change in current flow in a search coil, caused by imperfections in the sample. Both these methods can be used for the detection of surface and sub-surface micro-cavities as well as for the presence of slag particles or inclusions. The sensitivity of all the three methods decreases if the imperfections are spherical rather than elongated. For spherical micro-imperfections at or near the surface, optical or metallographic methods have to be employed.

For most cases, interpretation of the causes of imperfections, begins by examining the essential metallurgical phenomena of the process, most of which have been discussed in the previous chapters. For surface imperfections, the factors involved are surface cleanliness of liquid metals (Chapter 3), reactivity with the mould surroundings or gases (Chapters 2, 4 and 8), surface and bulk properties of the mould (Chapter 8), flow into moulds (Chapter 9) and solidification (Chapter 6). In many instances a metallographic examination may be sufficient to establish the identity and cause of imperfections; for example, sand or slag inclusions or gaseous cavities. However, it often happens, that an identical imperfection may be due to one of several causes, and additional physical or chemical identification methods

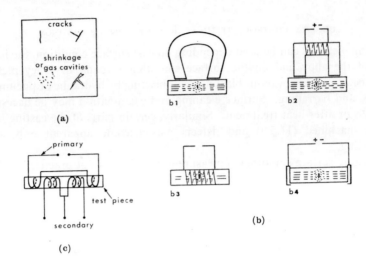

(a)

b1   b2

primary

test piece

secondary

b3   b4

(b)

(c)

Classification of surface defects by major groups		Origin of surface defects	
		Typical defect	Major origin
Cavities — Structure	segregation / indigenous inclusions	cracks, misrun, gas cavities, exudations, indigenous inclusions	alloy properties
Specific surface defects	cracks (tears) / holes (blowholes pinholes)	exogenous inclusions / cold shut, scab	liquid flow
	roughness, pitting	sink, warping, misrun	feeding
	scab, vein, crush / exogenous inclusions / sink / cold shut / burning, metal penetration	almost all specific surface defects, holes	mould
	misrun / flash / cross joint / warping	cracks, warping	casting design

(d)

FIGURE 11.3 Testing and metallurgical analysis of casting surface imperfections

(a) Penetrant dye markings.
(b) Magnetic flux testing methods: (1) permanent magnet, (2) electro-magnet, (3) coil and (4) current flow.
(c) Eddy current principle.
(d) Classification and origin of surface defects.

Each type of defect can be further subdivided according to size, shape, position, origin or other special features. Each may be associated with metal, mould, or casting process characteristics: for example, exogenous inclusions may be sand, slag, dross, or dressing; cracks may be micro or macro, hot or cold.

have to be applied. A still more complex, but frequent case is that where the full analysis of imperfections may require specially designed experiments based either on production or on laboratory scale and possibly interpreted by statistical analysis. A diagrammatic summary of the major types and likely origin of surface imperfections is shown in Fig. 11.3.

11.5.3 Size and dimensional accuracy of castings

Dimensional accuracy of a casting in a given alloy is primarily a function of the casting process, involving the pattern and mould equipment, the mould making method and the casting design. Contraction allowances for the pattern equipment are taken from data established empirically, and the main dimensional tolerances for a given casting are specified on the blue print. These are normally taken on the machined surfaces of a casting so that, for many applications of castings, the processes merely differ in the amount of machining required to achieve the required dimensions of the casting.

Provided that the chief casting process tolerances have been accurately arrived at, the main metallurgical problem in this field is the dimensional variation in production of identical castings. With mineral moulds, several factors can contribute towards the variation of mould cavity dimensions (e.g. mould rigidity, Chapter 8). Accurate positioning and rigidity of both mould parts and of cores in a casting process is very important. The second general source of variation in dimensions is that due to casting and residual stresses (11.5.6). Distortion and cracking of castings result from compositional changes in melting, particularly due to absorption of impurities, and fluctuations in the cooling rate or the degree of mechanical restraint of the mould components.

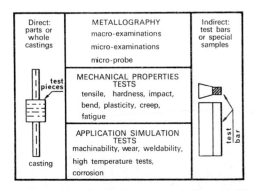

FIGURE 11.4   Metallurgical control charts

General cast structure control.   (*cont. over*)

	Process Controls (Quality)				Standard Specifications			
Control item	Variable	Sampling	Method	Control	Variable	Sampling	Method	Control
Alloy	composition	one or more stages during melting	chemical and physical analysis (Chap. 11)	record cards, graphs, statistical analysis	composition	before pouring	as process control	as process control
Melt quality	test bars, temperature, fracture, gas, fluidity, shrinkage	mainly before pouring or during melting	Chaps. 3, 4, 7 and 11	as with alloy composition	specified test bar or other tests	before pouring (sometimes attached to the casting)	Chap. 11	as process control
Metal moulds	temperatures, time, rates, pressures	periodic or continuous recording	metering by instruments	tables, graphs	none			
Mineral moulds	raw material, moulding mixture, gating and feeding elements	periodic	Chap. 8	record cards, graphs, statistical analysis	none			

Castings							
surface	almost all castings	Chap. 11	scrap, record cards, graphs, statistical analysis	as negotiated			
dimensions	all or representative castings	jigs and templates, Chap. 11	record cards	as negotiated			
structure (as-cast or heat-treated)	all or representative castings	Chap. 11	record cards, graphs, statistical analysis	properties dependent on structure	all or representative castings	Chap. 11	as process control
soundness (cavities)	all or representative castings	Chap. 11	record cards, graphs statistical analysis	cavities	all or representative castings	Chap. 11	as process control
stress	none or all castings	Chap. 11	record cards	as negotiated			

FIGURE 11.4 (*cont.*) Metallurgical control charts
Summary of metallurgical controls.

Dimensional accuracy and tolerance control is largely based on direct mechanical measurements on the finished castings, but, as subsidiary methods, eddy current (11.5.2) ultrasonic (11.5.5) pressure flow and other methods can be used.

11.5.4 Structure and structure-sensitive properties

The importance of cast structure in metallurgical control is largely dependent on the subsequent elastic and plastic property requirements. Plasticity behaviour is the governing factor in the structural control of most ingots. The most rigorous plasticity requirements are normally encountered in forging processes because of the non-uniform deformations involved, demanding cast structure of optimum plasticity (Chapter 7). Deformation conditions are less severe in rolling and still less so in extrusion. In all these cases the original cast structure is replaced in the finished products by a new structure resulting from recrystallisation or heat treatment. The structural control of ingots relies largely on the control of features in the structure which control ingot plasticity behaviour. With some alloys and ingots, cast structural features such as various types of segregation may persist in the final product, and these have to be controlled at the cast stage.

With a large group of shaped castings, particularly most pressure die castings, structural requirements are controlled by the alloy composition only. This implies that structural variations that may be encountered in the casting process, apart from those due to composition changes, do not adversely affect the acceptance standards properties of the castings. In the next group are those castings where the variation in the cast structure is insensitive to the casting process variables (apart from the composition) but the occurrence of structural defects, such as cavities or inclusions, does vary, necessitating some degree of structural control. Examples of this type are numerous, e.g., the deoxidation control of many alloys and sulphur control in several ferrous alloys. In the third group are castings where both the as-cast and the finally corrected structure have to be controlled, as with most of the heat-treatable alloys. The metallurgical control of structure in this case necessitates the control of the size, shape and distribution of some or all of the main constituents of the cast structure, as well as of 'unwanted' features such as inclusions and cavities.

In a few isolated examples of casting alloys non-destructive methods can be used for structural controls. For example, the volume ratio of non-magnetic constituents in the structure of a magnetic alloy can be measured, or in some alloys the characteristics of certain constituents, e.g. graphite in shape and size in grey iron, can be checked by ultrasonic methods (11.5.5). Bulk inclusions can also be detected by radiographic methods (11.5.5) and some surface or sub-surface inclusions by magnetic or eddy current methods (11.5.2). In the majority of cases, however, structural control implies a destructive method of examination, where a required number of production samples of castings, or specially cast samples such as test bars, are examined for various features of macro- or microstructure. Recent advances in both speed and resonance measurement of ultrasonic energy transmission through solids are likely to lead to considerable improvements in non-destructive methods of controlling the cast structure.

With some alloys and castings, structure-sensitive properties can be used for indirect structural control, such as hardness for grey iron castings, bend tests for spherolitic graphite and malleable irons, and tensile properties for steel and non-ferrous alloy castings. In general, the object of the structural control of castings is to achieve certain specific property standards. If these can be measured directly, they serve as a basis of structure quality control. But difficulties often arise with mechanical properties which may necessitate destruction of the casting. This is due to the fact that the structure varies not only in relation to a given cross-section but also from one part of the casting to another. A separately cast test bar can have similar properties to those of a particular casting section only, and cannot be used for general casting property evaluation. The alternatives are either to have a series of test bar sizes to typify various casting cross-sections, or to test directly various cross-sections of the casting itself, Fig. 11.4.

As a result of the essentially destructive nature of structural control methods for cast metals, two types of system can be applied in practice. In the direct system, whole cross-sections of castings, or smaller cut-out samples, are examined by the appropriate metallographic or other techniques. This system is readily applied to the structural control of ingots and in many instances to shaped castings. In the indirect system, cast samples, such as test bars, are examined. This system is based on the analogy of structure between the sample and the casting, and therefore has to be used with discretion (11.2.5). Such structural problems will be more satisfactorily resolved by the application of non-destructive methods which are being gradually developed at the present time.

11.5.5   Soundness of castings

Soundness generally receives more attention in the quality control of foundry processes than any other casting property. This is because, with chemical analyses and other controls, structural variations in castings can be maintained within the required limits, but unsoundness is often due to elements in the alloy not included in the specifications. In other cases, it can be due to independent causes, such as the feeding. Furthermore, unsoundness is far more likely to lower the property standards than other cast structural variations. The control of internal cavities in cast metals poses the same metallurgical problems as the surface and near-surface cavities considered in 11.5.2, i.e. detection and metallurgical interpretation.

The main methods used for examining the soundness of castings are non-destructive, Fig. 11.5, but to a far smaller extent direct destructive tests are also used. Sample or test bar indirect control can also be used provided that its limitations are recognised and understood, 11.2.5.

By far the most used methods for cavity detection are those based on various radiological techniques. Electro-magnetic energy waves of suitable characteristics, mainly X-rays and γ-rays, are passed through the casting and these produce, either on a fluorescent screen or, on a film, images or shadows of cavities. This effect is due to absorption by scattering of some of the waves passing through the structure of the casting. Those waves which encounter cavities in their path through the casting affect the screen or the film more

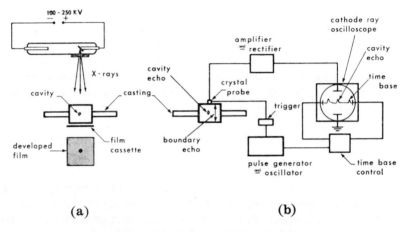

(a)   (b)

FIGURE 11.5 Cavity detection in castings

(a) X-ray radiography.
(b) Ultrasonic echo method.

strongly and the contrasts produced on the developed film or plate are evidence of the cavities present. The penetration of waves through the casting is controlled by the type of the waves used. Wave characteristics determine the choice of the radiological technique to be used for a given alloy and the range of casting thickness which can be examined. For example, with casting thickness variation from 1 to 7 in. of aluminium alloys, the required X-ray tube potential varies from 100 to 250 kV in order to obtain optimum film contrast. The same X-ray source would be suitable for steel thickness variation from 0·5 to 2·5 in. while γ-radiation from radioactive isotopes of various energies is usually selected for thicker castings in different alloys. Apart from this, the selection of γ-rays depends on the availability, cost and production control factors. With a given source and energy of radiation, the time of exposure and the nature of the film are also controlled to obtain the maximum contrast in detection. The size of the cavity that can be detected is of the order of 1% of the casting thickness. For cavities still smaller in size, microradiographic (destructive) methods have to be used. As some cavities can also be of an elongated kind, and because of the uncertainty of their orientation relative to the direction of the waves, it is frequently necessary to take radiographs in various directions through the same cross-section of a casting. The number of exposures to be taken on each casting or the number of castings to be examined from a batch depends on the engineering requirements grading of the casting. In general, radiological techniques are very costly and require a careful selection of suitable equipment, controlled installation of the equipment from the point of view of health hazards, and highest quality techniques in the taking and experience in the interpretation of radiographs.

Instead of electromagnetic energy waves, sound energy waves of suitable frequencies can be used for the detection and control of cavities. The frequencies used vary between 500 kc/s to 10 Mc/s, hence the term ultrasonic waves. The ultrasonic method has a number of different variants, some of which can be used for other structural or surface examinations, or casting thickness measurement. One of the most widely used principles is that of pulse or echo reflection, Fig. 11.5. A pulse of sound waves from the probe or transducer at the surface of a casting, e.g., a barium titanate crystal vibrating by applying alternating voltage to its face, is transmitted at regular intervals through the casting. The waves are reflected from a boundary, such as the opposite surface of the casting, or a cavity, and are received back by the same crystal, or by another crystal in the same probe, or by another probe. The boundary reflection is made visible by means of an oscillograph, Fig. 11.5, and hence any defect or cavity along the wave path will return the signal sooner, giving a new peak or 'blip' on the time-base of the oscillograph. In applying the ultrasonic method for cavity detection, the following main factors have to be taken into account: the frequency of the probe oscillator must increase with the decreasing size of cavities or the waves will be scattered instead of being reflected; the surface smoothness of the casting must be adequate for the probe to transmit the waves and a coupling medium such as grease, oil or water may be necessary; and the sensitivity of the oscillograph should be adjusted to eliminate spurious sound or boundary effects. It is clear that the optimum orientation of longitudinal cavities is perpendicular to the wave path. A typical minimum cavity size for detection lies above 0·001 in., mainly because at this order of size there is interference from normal features of the microstructure, such as grain boundaries. Both the structure of the material and the distance of the cavity from the surface, affect the numerical value of the minimum size cavity capable of detection.

Castings that are used for hydraulic or pressure applications are frequently tested for cavities by direct examination under pressure. The open ends of the casting are made pressure-tight by mechanical means, the casting is submerged into water and the required air pressure is applied inside the casting. Air leakage through the continuously connected cavities in the casting is thus made readily visible. For relatively small size castings, density measurement by Archimedes' principle can be used as a soundness control method. Finally, the macroscopic or microscopic metallographic methods for various cross-sections of representative castings can be used for the detection of cavities.

The application of cavity detection test results raises two separate problems; the importance of the effect of cavities on the subsequent properties of the casting, and the explanation of the origin of the cavities so that the casting techniques can be modified to improve the soundness of the casting. In ingots, cavities may affect the deformation behaviour of an alloy or the final properties of the product. A particular behaviour is determined by the intrinsic plasticity of the alloy, the method of plastic deformation used and the nature of the surfaces enclosing the cavity. For many applications of shaped castings the presence of some external or internal cavities is not detrimental. In such cases, faulty castings can be 'reclaimed' by various processes of impregnation to close up the surface holes. Various welding processes are also widely used for the reclamation and repair of castings. The more critical problem is that of the effect of residual cavities on the application of castings for stressed components. In this field cavity detection tests are subject to various rules and regulations of industrial and governmental organisations, largely based on empirical data and summarised in standard specifications.

The origin of cavities can normally be explained by either gas (Chapters 4 and 8) or volumetric (Chapter 10) contraction effects. In many instances the full production quality control data may be sufficient to account for the actual causes of cavities. Cavities due to mould gas blowing or due to hot spot shrinkage are readily identified by their location, shape and surface conditions. In many instances, however, statistical analyses of process control data may have to be supported by additional production or laboratory experiments to achieve a complete understanding of the cause of the cavities.

11.5.6 Stresses in castings

Most castings solidify under non-equilibrium conditions and therefore in principle must contain some residual internal stress. If such castings are subsequently heat-treated for structural improvement in mechanical properties, then the internal stresses are relieved at the same time to 'safe limits'. Other castings may be specifically heat-treated for stress relief. The large majority of castings are used in the original internally stressed condition. Present industrial production control of stresses in castings thus operates on three principles:

(a) The stresses generated are kept below the level leading to either hot or cold cracks in the castings, by controlling the various sources of stress.

(b) The residual stresses may be left in the casting or relieved by heat treatment, depending on the application of the casting and the level of stress.

(c) Stress magnitude measurement is not normally carried out in either (a) or (b), owing to the experimental complexity of the methods of stress measurement at present available.

The simplest principle which can be applied to stress determination is that based on the measurement of the strain produced by cutting through the stressed casting. In the case of a ring casting, opening up of the ring after cutting can be used to calculate the stress. With ingots or with shaped castings, partial machining of the casting surface unbalances the total stress and leads to a strain which can be measured, for example, by a strain gauge. The residual stress pattern which existed in a casting can be obtained in this way. These and other methods, such as those based on X-rays, do not readily lend themselves to process control. A general summary of process and specification controls is given in Fig. 11.4.

11.6 SUMMARY

Founding is basically a human activity, hence the importance of process and products controls. Indeed, it is in this field that the whole process acquires a concept of unity, the blue print marking the beginning and the performance of the casting the end of the process quality control. It is in this field too that the degree of co-operation between the engineer, the metallurgist and the production engineer receives its crucial test. Formulating, operating and improving the quality control system and methods is one of the keys to the economic progress and to the future growth of founding as an industrial process.

11.7 GENERAL SUMMARY: FOUNDING—ENGINEERING, SKILL OR APPLIED SCIENCE?

The general theme of this book has been to analyse the behaviour of alloys through the various stages of casting and to attempt to explain this behaviour in terms of elementary scientific principles. It is clear from this analysis that many castings have been and are likely to be made in future on the basis of skill and experience. It is also evident that in other cases all the prior problems can be solved by applied scientists or engineers, the process in the end becoming fully rationalised. One of the main tasks, therefore, for a young student thinking of entering the foundry industry is to consider with equal appraisal and understanding the merits and limitations of both the skilled and the engineering or applied science approaches to casting processes. Unacceptable castings can be produced with too much skill or too much science, but more frequently when both are lacking.

FURTHER READING

BURBRIDGE, J. L., *The Principles of Production Control*, Macdonald & Evans, London, 1962.
SCHROCK, E. M., *Quality Control and Statistical Methods*, Reinhold, New York, 1957.
CHALMERS, B., and QUARRELL, A. J., *The Physical Examination of Metals*, Arnold, London, 1960.
Statistical Quality Control for Foundries, American Foundrymen's Society, 1953.
HANSTOCK, F. C., *The Non-Destructive Testing of Metals*, Institute of Metals, 1951.
HOGARTH, C. A. and BLITZ, J., *Technique of Non-Destructive Testing*, Butterworth, London, 1960.
Analyses of Casting Defects, American Foundrymen's Society, 1947.
Atlas of Defects in Castings, Institute of British Foundrymen, 1961.
ARMSTRONG, W. H., *Mechanical Inspection*, McGraw-Hill, 1953.
ANGUS, H. T., *Physical and Engineering Properties of Cast Irons*, British Cast Iron Research Association, 1960.
British Standard Mechanical Tests for Metals, British Standards Institution, 1951.
EVEREST, A. B., *Br. Foundryman*, 1964, **57**, 273.

INDEX